Modelling the Deformation, Recrystallization and Microstructure-Related Properties in Metals

Editor

Jurij J. Sidor

MDPI • Basel • Beijing • Wuhan • Barcelona • Belgrade • Manchester • Tokyo • Cluj • Tianjin

Editor
Jurij J. Sidor
Savaria Institute of Technology,
Faculty of Informatics,
Eötvös Loránd University
Hungary

Editorial Office
MDPI
St. Alban-Anlage 66
4052 Basel, Switzerland

This is a reprint of articles from the Special Issue published online in the open access journal *Metals* (ISSN 2075-4701) (available at: https://www.mdpi.com/journal/metals/special_issues/Modelling_Deformation_Recrystallization).

For citation purposes, cite each article independently as indicated on the article page online and as indicated below:

LastName, A.A.; LastName, B.B.; LastName, C.C. Article Title. *Journal Name* **Year**, *Volume Number*, Page Range.

ISBN 978-3-0365-2384-2 (Hbk)
ISBN 978-3-0365-2385-9 (PDF)

Contents

About the Editor

Jurij J. Sidor is full professor at Eötvös Loránd University (ELTE), Faculty of Informatics, Savaria Institute of Technology in Hungary. He gained his PhD in the field of Materials Science in 2004, habilitation as associate professor in Materials Science and Technology in 2017, and defended the doctoral thesis for gaining the academic title of Doctor of Sciences (DrSc) in 2021. In his research career, he has been working at the Institute of Materials Research of the Slovak Academy of Sciences (Slovakia), Delft University of Technology (Netherlands), Ghent University (Belgium), University of Science and Technology of Beijing (China) and University of West Hungary (Hungary). The research activities are related to the development of both novel processing technologies and quantitative physically based models. These numerical approaches help the optimization of production technology and materials properties. The research area of the author covers the field of virtual material processing, materials characterization by Scanning Electron Microscopy, electron backscatter diffraction and mechanical testing. In this field, Prof. Sidor published more than 70 scientific papers, two book chapters and presented his results at numerous international scientific events as keynote or invited lecturer. According to Scopus and Google Scholar databases, his scientific publications have been cited more than 1000 times. Numerous discoveries in the fields of modelling and examination of microstructure and texture evolution in polycrystalline metallic systems resulted in several national and international awards and recognitions. For his reviewing activity, he was awarded several times by outstanding reviewer's recognition.

Preface to "Modelling the Deformation, Recrystallization and Microstructure-Related Properties in Metals"

Experimental investigations of the thermomechanical processing (TMP) of metals clearly demonstrate that technological process tuning parameters have a great influence on both the microstructure and texture evolution, which determine the chemical, physical, or mechanical properties. During the processing chain, the behavior of polycrystalline material is correlated with the grain size, grain crystal structure, and crystallographic orientation. The mesoscopic transformations of polycrystalline aggregates, involving microstructural and crystallographic changes on the grain level, can be interpreted by a vast body of modelling approaches developed. Advances in modelling have created a solid platform for understanding the evolution of microstructural features in polycrystalline systems during particular processing steps and bring to light many hidden aspects of production as well as assist in revealing the behavior of materials under specific circumstances. Since mesoscopic changes in TMP are "genetically" connected, the modelling techniques enable tuning particular processing step to tailor the desired material performance for a given application.

In this Special Issue, we intend to provide a wide spectrum of articles dealing with the modelling of microstructural changes in various metallic systems. The major aim was to discuss different features of microstructure evolution such as morphological changes, crystallographic aspects, phase transformation as well as structure-property relationships in polycrystalline aggregates.

Jurij J. Sidor
Editor

metals

Editorial

Modelling the Deformation, Recrystallization, and Microstructure-Related Properties in Metals

Jurij J. Sidor

Savaria Institute of Technology, Faculty of Informatics, Eötvös Loránd University (ELTE), Károlyi Gáspár tér 4, 9700 Szombathely, Hungary; js@inf.elte.hu

1. Introduction and Scope

Experimental investigations of the thermomechanical processing (TMP) of metals clearly demonstrate that technological process-tuning parameters have a great influence on the evolution of both microstructure and texture, which determine the chemical, physical, and mechanical properties of metals. During the processing chain, the behavior of polycrystalline material is correlated with the grain size, grain crystal structure, and crystallographic orientation. The mesoscopic transformations of polycrystalline aggregates, involving microstructural and crystallographic changes on the grain level, can be interpreted using the vast body of modeling approaches that have been developed. Advances in modeling have created a solid platform for understanding the evolution of microstructural features in polycrystalline systems during particular processing steps and bring to light many hidden aspects of production, as well as assisting in revealing the behavior of materials under specific circumstances. Since mesoscopic changes in TMP are "genetically" connected, modeling techniques enable the tuning of a particular processing step to tailor the desired material performance for a given application.

In this Special Issue, we provide a wide spectrum of articles dealing with the modeling of microstructural changes in various metallic systems. The main aim is to discuss different features of microstructure evolution, such as morphological changes, crystallographic aspects, and phase transformation, as well as structure–property relationships in polycrystalline aggregates.

Citation: Sidor, J.J. Modelling the Deformation, Recrystallization, and Microstructure-Related Properties in Metals. *Metals* **2021**, *11*, 1759. https://doi.org/10.3390/met11111759

Received: 20 October 2021
Accepted: 29 October 2021
Published: 2 November 2021

2. Contributions

The exceptional properties of various metallic polycrystalline systems, such as high strength, excessive hardness, great ductility at room temperature, superior energy absorption capacity, and good corrosion resistance, make them outstanding candidates for a wide variety of applications where one or another of the mentioned qualities, or the combination of several, is of crucial importance. The presented contributions [1–9] evidently demonstrate that the properties of metallic materials are microstructure-dependent and, therefore, the thermomechanical processing (TMP) of the polycrystalline aggregates should be strictly controlled to guarantee the attainment of the desired suite of qualities. Given this, the assessment of microstructure evolution in metallic systems is of extraordinary importance. Since the trial–error approach is a time-consuming and rather expensive methodology, the materials research community tends to employ a wide spectrum of computational approaches to simulate each chain of TMP and tune the processing variables to ensure the necessary microstructural state for achieving the desired performance in the final product.

In the most general case, the TMP of metals involves a sequence of deformation and annealing processes (or the combination of both into one technological step) that lead to morphological changes in the polycrystalline aggregate, the evolution of crystallographic texture, or phase transformation. All these aspects of microstructure development are partially discussed in the presented contributions [1–9]. In the first manuscript [1], the deformation flow in a single-phase alloy with a face-centered crystal structure is analyzed by means of a finite element (FEM) approach and computationally efficient flow-line model (FLM), which enabled the effective simulation of cold rolling in terms of the evolution of crystallographic texture. It was shown that by using the flow-line model, coupled with a Taylor-type homogenization crystal plasticity model, it is possible to carry out a texture simulation that is close to the one performed with the deformation history obtained by means of computationally expensive FEM. In addition, the correlation between the FLM model parameters and the rolling process quantitative indicators was defined, implying that the employed analytical approach was efficient in performing simulations of cold rolling without fitting constraints.

The second contribution [2] likewise addresses the crystallographic aspect of deformation by employing a well-established viscoplastic, self-consistent (VPSC) crystal plasticity approach. The numerical investigation is related to the deformation mechanisms in a magnesium alloy with a hexagonal crystal structure, where the polycrystalline system was subjected to uniaxial tension and compression. The influence of slip systems on the plastic deformation, mechanical response, and texture evolution was analyzed by the VPSC via engaging various slip and twinning activities.

The annealing phenomena are addressed in the third paper [3], where the authors investigated the phenomenon of dynamic recrystallization in a FeMnSiCrNi shape memory alloy subjected to hot compression. The analysis is based on a simulation of recrystallization using a cellular automaton approach, and apart from the microstructural features, both flow stress and dislocation density were predicted.

The latest advances in the modeling of mechanical performance based on microstructural features of polycrystalline systems are described in five contributions [4–8], which reveal the behavior of different materials under diverse processing operations. The springback phenomenon was simulated for the hot-stamping process in [4]. The calculations rely on the analysis of microstructures that evolved in the investigated boron steel. The relationship developed in the frame of the contribution [4] enables the estimation of the springback angle based on the area fraction of microstructures that evolved during the quenching. The following paper [5] deals with the analysis of the hot workability of 300 M steel, where the relation between the flow stress and microstructure evolution was examined. The investigation was conducted via in situ and ex situ compression tests. In another manuscript [6] equally concerned with hot deformation, the double-open multidirectional forging process is analyzed. In this investigation, the authors present a constitutive equation for GH4169 superalloy and simulate the microstructure evolution during dynamic recrystallization using the finite element approach. Anisotropy of plastic yielding is described in [7], where the authors investigate the mechanical response of the cross-rolled aluminum sheets. The simulation of cup earing is based on the evolved crystallographic texture. The model employed for the simulation of anisotropic behavior takes texture intensities into account, which appear in the {h00} pole figures. The performance of monocrystalline Cu-13.7% Al-4.2% Ni alloy is characterized in [8]. In that contribution, the authors investigate the so-called reversible martensitic transformation and describe both physical and structural changes in the complex metallic system induced by the thermocyclic treatments under applied loads.

Phase transformation is studied in [9]. In this study, high-strength steel was subjected to a hot-stamping process, and phase transformation was investigated via simulation of the transformation during the pressure-holding quenching process. The kinetics of phase alteration, which leads to the formation of ferrite, pearlite, and bainite, was analyzed according to the Kirkaldy–Venugopalan model.

3. Conclusions and Outlook

In this Special Issue, we present a wide spectrum of articles dealing with the modeling of microstructural aspects involved in deformation and recrystallization as well as simulation of microstructure-based and texture-based properties in various metals. The latest advances in the theoretical interpretation of mesoscopic transformations based on experimental observations are also partially discussed. The studies dealing with the modeling of structure–property relationships are likewise analyzed in the present collection of manuscripts.

Although, in the submitted works, many hidden facets of various technological processes and related microstructural changes were revealed by employing advanced computational approaches, the contributions collected in this issue nevertheless clearly show that further efforts are required in the field of modeling to understand the complexity of the material's world. The ultimate goal of modeling efforts is, arguably, the development of a comprehensive model that is capable of describing many aspects of microstructure evolution during thermomechanical processing.

Funding: This research was funded by the EFOP-3.6.1-16-2016-00018 project: "Improving the role of research + development + innovation in higher education through institutional developments assisting intelligent specialization in Sopron and Szombathely". The work was performed in the frame of the Széchenyi 2020 program: "Innovative processing technologies, applications of energy engineering, and implementation of wide-ranging techniques for microstructure investigation".

Conflicts of Interest: The author declares no conflict of interest.

References

1. Sidor, J. Assessment of Flow-Line Model in Rolling Texture Simulations. *Metals* **2019**, *9*, 1098. [CrossRef]
2. Lian, Y.; Hu, L.; Zhou, T.; Yang, M.; Zhang, J. Numerical Investigation of Secondary Deformation Mechanisms on Plastic Deformation of AZ31 Magnesium Alloy Using Viscoplastic Self-Consistent Model. *Metals* **2019**, *9*, 41. [CrossRef]
3. Wang, Y.; Xing, X.; Zhang, Y.; Jiang, S. Investigation of the Dynamic Recrystallization of FeMnSiCrNi Shape Memory Alloy under Hot Compression Based on Cellular Automaton. *Metals* **2019**, *9*, 469. [CrossRef]
4. Li, X.; Yan, X.; Zhang, Z. Springback Prediction of a Hot Stamping Component Based on the Area Fractions of Phases. *Metals* **2019**, *9*, 694. [CrossRef]
5. Chen, R.; Xiao, H.; Wang, M.; Li, J. Hot Workability of 300M Steel Investigated by In Situ and Ex Situ Compression Tests. *Metals* **2019**, *9*, 880. [CrossRef]
6. Jin, Y.; Xue, H.; Yang, Z.; Zhang, L.; Zhang, C.; Wang, S.; Luo, J. Constitutive Equation of GH4169 Superalloy and Microstructure Evolution Simulation of Double-Open Multidirectional Forging. *Metals* **2019**, *9*, 1146. [CrossRef]
7. Benke, M.; Schweitzer, B.; Hlavacs, A.; Mertinger, V. Prediction of Earing of Cross-Rolled Al Sheets from {h00} Pole Figures. *Metals* **2020**, *10*, 192. [CrossRef]
8. Matlakhova, L.; Pereira, E.; Pulnev, S.; Shigue, C.; Palii, N. Physical and Structural Characterization of Monocrystalline Cu-13.7% Al-4.2% Ni Alloy Submitted to Thermo-Cyclical Treatments under Applied Loads. *Metals* **2020**, *10*, 219. [CrossRef]
9. Zhang, P.; Zhu, L.; Xi, C.; Luo, J. Study on Phase Transformation in Hot Stamping Process of USIBOR® 1500 High-Strength Steel. *Metals* **2019**, *9*, 1119. [CrossRef]

metals

MDPI

Article

Assessment of Flow-Line Model in Rolling Texture Simulations

Jurij J. Sidor

Savaria Institute of Technology, Faculty of Informatics, Eötvös Loránd University (ELTE), Károlyi Gáspár tér 4, 9700 Szombathely, Hungary; js@inf.elte.hu; Tel.: +36-9450-4318

Received: 26 September 2019; Accepted: 10 October 2019; Published: 12 October 2019

Abstract: The nature of the thermomechanical processing of materials can be revealed by means of various numerical approaches. The accuracy of a particular model is linked to the boundary conditions employed. Intensive research activities over the past several decades in the field of finite element modeling (FEM) have enabled the development of various processing chains for particular purposes; however, this technique is computationally expensive, and in many instances, the behavior of materials during a processing step is analyzed by highly efficient analytical models. This contribution focuses on the implementation of a recently developed flow-line model (FLM), which enables the effective texture simulation of cold rolling. The results of numerous calculations, performed for a wide spectrum of roll gap geometries and various friction conditions, revealed that the deformation history predicted by the FLM employed was comparable to FEM calculations. A correlation was defined between the FLM model parameters and the rolling process quantitative indicators, implying that this analytical approach is capable of performing simulations of cold rolling without fitting constraints. It was shown that FLM coupled with a Taylor-type homogenization crystal plasticity model (Alamel) could carry out a texture simulation close to the one performed with deformation history obtained by means of FEM.

Keywords: cold rolling; deformation flow; texture simulation

1. Introduction

Metals are polycrystalline aggregates composed of numerous grains that can be considered perfect crystals if the materials are fully recrystallized. In order to predict the behavior of polycrystalline materials subjected to macroscopic load, the relationship between microstructures and properties should be known, and therefore, physically based modeling is of key importance. The understanding of microstructural changes is particularly important in setting up the thermomechanical processing (TMP) route. In TMP, the as-cast microstructure is subjected to continuous microstructural transformation. In the case of flat-rolled products, such as Al sheets, this transformation is conditioned by both deformation and recrystallization phenomena, since after casting, the material is subjected to hot rolling, which is followed by cold rolling, and finally the sheet is recrystallized during the final annealing process.

Various modeling approaches that employ diverse length scales (Figure 1) have been developed to reveal the behavior of materials during TMP [1] It is a common practice to employ phenomenological models for the optimization of properties in groups of metals that do not reveal substantial varieties in chemical composition [1,2]. This strategy likewise works for tailoring a certain metal grade for a particular application. The main disadvantage of phenomenological approximations is that the model parameters need to be recalibrated in the case of a metal grade change or if the TMP quantitative indicators are varied in a wider range of processing windows. In more advanced numerical formulations, such as finite element modeling (FEM), materials are considered as continua, and thus, the mesoscopic phenomena, occurring on the grain levels, are not accounted for in this engineering length scale (Figure 1). Nevertheless, FEM is successfully employed in many fields, especially when there is a need to simulate the response of a system subjected to mechanical load. On the other end of the length-scale spectrum, macroscopic properties are derived from interatomic potentials and by engaging basic principles of molecular thermodynamics, which ensures generic information about the relationships between atomic arrangements and macroscopic properties; however, this formulation is applicable only for ideal systems. The application of atomistic length-scale simulations to metals is restricted due to limitations in computational resources and also because of the fact that polycrystalline structures contain too many crystal defects even in a fully recrystallized state (typically 10^8 m^{-2}).

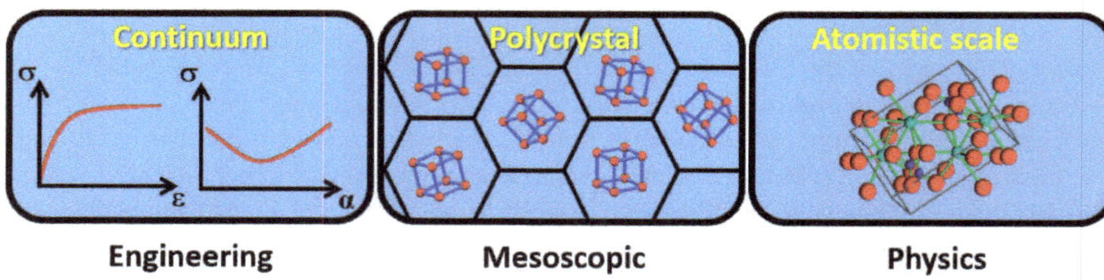

Figure 1. Variety of length scales involved in modeling the microstructure–properties relationship.

Since a 1-mm-thick metallic sheet (A4 size (21 cm × 29.7 cm)) contains approximately 4 billion grains of various crystal orientations, assuming that the average grain size is 25 μm, it becomes clear that neither engineering (continuum) nor atomistic scales are capable of revealing the true nature of mesoscopic transformation involved in TMP. To capture the essence of microstructural changes in industrially produced materials, both mean-field and full-field crystal plasticity (CP) models are employed [1,3–9]. Either approach implements a certain homogenization scheme insofar as performing calculations for the above-mentioned 4 billion crystals, consisting of even smaller entities, would require enormous computational power. In CP, homogenization implies a characteristic length scale (mesoscopic) represented by grains of polycrystalline aggregate, and these entities are treated as perfect crystals (the in-grain heterogeneities are usually not taken into account). The mesoscopic scale tends to fall between two ends of the characteristic length spectrum, and this level of order is important from a practical perspective, as many of the known properties are controlled by the grain size and crystallographic texture. CP approaches treat the polycrystalline material as a continuum at the level of the crystal because the grain size is order of magnitudes larger compared to the size of molecular domains.

Full-field models [1,3,4] allow for analyzing not only the microscopic and macroscopic responses of material but also account for the evolution of heterogeneities at both levels. Although these methods give rise to a more comprehensive representation of a microstructure and provide detailed information on the deformation flow on meso- and macrolength scales, their practical implementation is limited because of extensive computational time. In contrast, the mean-field models are more computationally efficient, but they have fewer degrees of freedom compared to the full-field methods. For instance, in mean-field approximations [5–9], among other things, each grain interacts with a neighboring one in a way defined by the homogenization scheme, and the misorientation in the grain developed during deformation cannot be captured. Even though some microstructural features are ignored in Taylor-type homogenization CP formulations (such as "visco-plastic self-consistent (VPSC) [7], advanced Lamel (Alamel) [5,6], Cluster V [8], etc.), simulations on a mesoscopic scale are still capable of providing generic knowledge on the relationship between the structure, properties, and performance of a polycrystalline aggregate. Polycrystal models have been successfully employed in simulating the evolution of texture in metals (subjected to diverse strain modes), calculating the plastic strain ratio as well as predicting the cup-earing [5–11].

Among the many homogenization theories, Alamel [5,6] was employed in this study to simulate the texture evolution in an Al alloy. In this mean-field approach, the grain assembly is subdivided into a number of clusters consisting of two grains, whereas the interaction inside the cluster is governed by the randomly chosen grain boundary orientation. For a given pair of grains, the equilibrium of stresses is accomplished via the strain compatibility. Similarly to the above-listed Taylor-type models [5–9], the grain pair is subjected to a strain mode identical to the macroscopic one. In crystal plasticity formulations (including Alamel), the evolution of texture in face centered cubic (FCC) metals during cold deformation is related to the crystallographic slip on octahedral slip systems, whereas the macroscopic strain mode is approximated by various models. In order to make the application of crystal plasticity modeling practically attainable, the CP should be coupled with the computationally effective approach, which is capable of accurate prediction of deformation flow in a material. When it comes to rolling, the simplest approximation is the plane strain compression, which disregards many aspects of the process. Because of its simplicity, this strain mode is often used in simulations and tends to provide a reasonable estimate of overall (bulk) texture evolution [5,6,8,10]. In FEM, the effect of many technological parameters on the deformation flow can be taken into account, while the displacement history is accurately calculated for various materials at diverse temperatures; however, the model set-up and the calculation procedure are time-consuming. In view of this, deformation is often approximated by analytical solutions [12–18] such as flow-line models, which are capable of capturing many aspects of the process; however, the practical implementations of these computationally effective approaches are limited by fitting parameters, which have to be derived from the experimental data for each particular case.

Modeling the evolution of deformation texture in Al alloys by combining the basic principles of CP theories and models that are capable of predicting deformation flow is vital, since many crystallographic features evolved in deformation are directly linked to the evolution of recrystallization texture [19–21]. The latter determines plastic anisotropy and limits the forming characteristics of a material.

Even though many efforts have been made to model texture development during cold rolling, there are nevertheless many aspects of texture evolution that are still not explained by existing models. The goal of this contribution was to employ computationally efficient and accurate numerical approaches that could ensure quantitatively reasonable texture simulations for the cold rolling process. For this, a well-established CP model (Alamel) [5,6] was coupled with the recently developed FLM [14]. To make the CP simulations practically applicable, a correlation was defined between the FLM model parameters, the roll gap geometry, and the friction coefficient. This implies that combined Alamel–FLM-based modeling can be performed without a fitting procedure, allowing for analyses of the effect of technological quantitative indicators on texture evolution in cold rolling. It is shown that the Alamel–FLM texture simulations revealed both qualitative and quantitative resemblance to the counterparts calculated with the deformation history obtained by means of finite element modeling. The quality of the texture simulations was estimated by comparing the simulated textures to the experimentally measured ones.

2. Materials and Computational Methods

In the current investigation, the evolution of crystallographic texture was studied in an Al–Mg–Si Al alloy from 6016 series. The sheet (1.125 mm thick) was subjected to a 29.6% thickness reduction in a single pass with a roll diameter of 128.9 mm. Prior to cold rolling, the material was annealed at 550 °C to ensure a fully recrystallized microstructure. Both recrystallized (initial) and deformed (rolled) samples were subjected to through-thickness texture measurements by means of the electron backscattering diffraction (EBSD) technique. The orientation data were collected and analyzed by commercial OIM-TSL-8® software (EDAX Inc., Mahwah, NJ, USA).

Sample preparation for orientation imaging microscopy (OIM) was performed according to the standard procedure, which comprises mechanical grinding and polishing as well as electrolytic polishing. The mechanical polishing procedure was finished with DiaDuo-2 Struers®-type 1-μm diamond paste. Electrolytic polishing, as a final step of sample preparation for EBSD, was conducted for ~1 min at a voltage of 18 V with A2 Struers® electrolyte and was cooled to temperatures ranging between −5 and 0 °C.

A Hikari-type EBSD detector was attached to a scanning electron microscope (SEM) FEI®. The OIM data of the investigated material with a fully recrystallized microstructure were collected at an acceleration voltage of 20 kV, whereas in the cold-rolled sample, to avoid overlap of the acquired pattern with ones originating from the deeper layers, the acquisition was performed at a comparatively low acceleration voltage of SEM. An application of 15 kV guaranteed appropriate OIM data acquisition. During OIM measurements, the sample was 70° tilted with respect to the EBSD detector. EBSD mapping was performed on a hexagonal scan grid in the plane perpendicular to the sample transverse direction (TD-plane) extending over the entire thickness of the investigated sample. For accurate texture evaluation, the deformed sample was scanned at a step size of 1.5 µm, while the recrystallized material was investigated at a 5-µm step. In order to ensure a meaningful comparison between the experimentally measured and simulated orientation distribution functions (ODFs), the data were postprocessed with MTM-FHM software [22], and the textures were displayed in φ_2 = const. sections.

The evolution of the cold-rolling texture was analyzed by a well-established Taylor-type homogenization approach called Alamel, the comprehensive formulation of which can be found elsewhere [5,6]. The simulations were performed by taking into account {111}<110> octahedral slip systems, which are typically activated in FCC metals during cold-forming processes. Prior to texture simulations, a homogenization of experimentally measured EBSD data, which involves the determination of the representative volume element, had to be performed. In the current case, the measured OIM data were converted to a continuous ODF and afterwards discretized to a set of 8000 equally weighted orientations.

To enable the calculation of texture evolution through the Taylor-type homogenization scheme employed, the deformation history in the form of strain velocity gradients should be provided to the given CP model. Both a two-dimensional finite element method (Deform 2D®) and a recently developed flow-line model (FLM) [14] were employed in the calculation of deformation flow. These approximations are capable of capturing strain heterogeneities across the thickness of a cold-rolled sheet. In the FLM and FEM simulations, the material subjected to rolling is considered to be plastic and isotropic. In FEM, the strain hardening phenomenon is described with a stress–strain curve fitted by piecewise linear segments, whereas the rolls are considered to be fully rigid objects. Contrary to FEM, the FLM employed neglects the strain hardening effect. The following material parameters were used in the FEM simulations for the isotropic aluminum matrix: E = 68.9 GPa (Young's modulus), ν = 0.33 (Poisson's ratio), and σ_y = 80 MPa (yield stress). Since rolling is a symmetric deformation process, the deformation flow was simulated for the following layers: surface, 1/5, 2/5, 3/5, and 4/5 of the half-thickness and midthickness plane. The extracted outputs were used as an input for CP calculations. The results obtained for the FEM-CP and FLM-CP simulations were compared to each other and likewise to their experimentally measured counterparts.

3. Results

3.1. Brief Description of the Flow-Line Modeling Approach

The flow behavior of a material subjected to rolling can be described by various approaches with a diverse degree of accuracy. In many instances, rolling is considered to be the plane strain compression deformation, since this abstraction enables the fast estimation of strain. In this approach, the strain mode is homogeneous across the thickness, while both the strain path changes and the evolution of shear components, caused by friction and roll gap geometry, are neglected. Numerous analytical approaches have been developed for a more accurate description of materials' behavior in deformation, and the main advantage of these models [12–18] over other numerical approaches, such as finite element modeling [1], is their efficiency and accuracy. Flow-line models [12–18] are analytical approaches that are capable of describing the deformation stream under given boundary conditions for a particular process. These formulations have been intensively employed for rolling force and torque calculations and the design of asymmetric and vertical rolling processes [12,13,15–18]. In the FLM employed, the detailed mathematical description of which is described elsewhere [14], a kinematically admissible displacement velocity field fulfills the following boundary conditions: (a) the entrance and the exit velocities of a rolled sheet are even across the thickness; (b) the incompressibility condition is fulfilled at all points; (c) a material's flow occurs along the prescribed streamlines; (d) at the surface, the velocity field is conditioned by means of model parameter α, which guarantees a difference between the velocities of the surface and midthickness layers; (e) the variation of the velocity across the thickness is conditioned by the second model parameter n; and (f) the approximation does not allow for any displacement in the transverse direction (TD).

In the FLM employed [14], the material's flow occurs along the prescribed streamlines, determined by means of Equation (1),

$$z_s = \phi(x,z) = z\varsigma(x) = \begin{cases} z\frac{1}{e}\left[1+\left(\frac{s}{e}+\left(1-\frac{s}{e}\right)\left(\frac{L_d-x}{L_d}\right)^{2.1}\right)^{-m}\right]^{\frac{1}{m}} & x \le L_d \\ z\frac{1}{e}\left[1+\left(\frac{s}{e}\right)^{-m}\right]^{\frac{1}{m}} & x > L_d \end{cases} \quad (1)$$

with projected contact length (L_d)

$$L_d = R\sin(\theta) \quad (2)$$

and

$$\theta = \cos^{-1}([R + s - e]/R), \quad (3)$$

where θ is a bite angle and R is a roll radius; e is the half-thickness of the sheet prior to rolling; s is the half-thickness of the sheet after rolling (see Figure 2 for details); and z_s corresponds to the position of the flow-line ($0 \le z_s \le 1$, $z_s = 0$ for the midthickness and $z_s = 1$ for the surface layer).

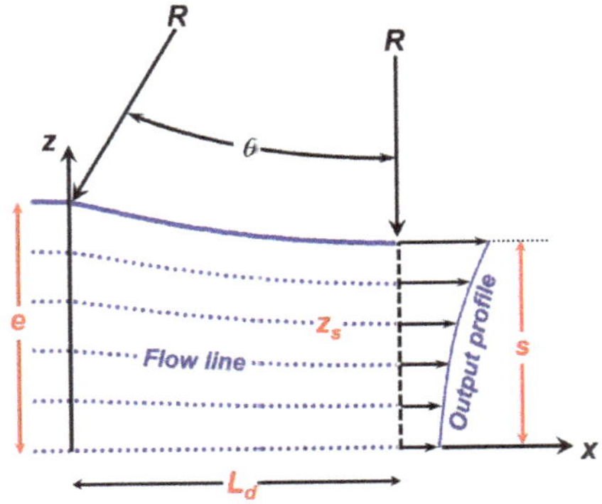

Figure 2. Schematic illustration of sheet geometry in a roll gap with the parameters of the flow-line model employed [14].

In this analytical approach, the value of m was set to 100 with the aim of ensuring a continuous first and second derivate of the $\varsigma(x)$ function in Equation (1) at $x = 0$. In the model developed by Decroos et al., the exponent of 2.1 (see Equation (1)) guarantees the following: (i) a quasi-parabolic shape of $\varsigma(x)$ in the deformation zone and (ii) continuous first and second derivatives of this function when $x = L_d$. Both m and the exponent of 2.1 ensures various profiles for streamline z_s, depending on the roll gap geometry. If $z_s = 0$, the deformation flow occurs along the straight streamline, while for $z_s = 1$, the flow is conditioned by the roll radius.

Both the entry and the exit velocities of a rolled sheet are even across the thickness; however, the shape of the output profile (see Figure 2) is controlled by two model parameters (α and n), which are supposed to be positive float numbers. Within a deformation zone, the model parameter α guarantees a difference between the x-component velocity v_x of the surface and midthickness layers. The variation in velocity across the thickness (z-component) is conditioned by the second model parameter n. As Figure 3 reveals, for a given value of n, v_x is identical for all layers with various z_s if $\alpha = 0$, while the difference between the velocities along the flow-lines tends to rise when α follows an upwards trend. Increasing the value of n in the FLM model while α remains constant tends to significantly enhance the velocity v_x of the subsurface layers with respect to the midthickness planes. In the case of $\alpha > 0$, the v_x of all through-thickness layers tends to converge to a single point at x corresponding to a neutral point N. From this point onward, the flow of the midlayer is faster than the top one. The position of the neutral point φ_n in rolling can be computed by the following equations [2]:

$$\phi_n = \sqrt{\frac{h_f}{R}}\tan\left(\sqrt{\frac{h_f}{R}}\frac{H_n}{2}\right), \tag{4}$$

$$H_n = 0.5\left(H_0 - \frac{1}{\mu}\ln\left(\frac{h_i}{h_f}\right)\right), \tag{5}$$

$$H_0 = 2\sqrt{\frac{R}{h_f}}\tan^{-1}\left(\sqrt{\frac{h_i - h_f}{h_f}}\right), \tag{6}$$

$$L_{dN} = L_d - R\phi_n, \tag{7}$$

where μ is a friction coefficient, and L_{dN} is the x component of a neutral point.

Although v_x is identical for all z_s when $\alpha = 0$, the z-component of velocity (v_z) reveals significant deviations, implying that surface layers experience higher straining levels compared to the midthickness plane ($z_s = 0$). It should be noted that even if $\alpha = 0$, the strain path of a rolled sheet is different from the plane strain compression due to displacement heterogeneities caused by various v_z across the thickness. For a given value of α, the effect of n on v_z seems to be negligibly small; however, the same cannot be said about the deformation patterns revealed in Figure 4 (see cases for $\alpha = 1.5$, $n = 1.5$; and $\alpha = 1.5$, $n = 2.5$). It can be concluded from Figures 3 and 4 that employing diverse α and n parameters allows for reproducing a wide spectrum of deformation patterns obtained under different thermomechanical processing conditions.

This computationally efficient model enables the fast calculation of deformation velocity gradients evolved across the thickness of a rolled sheet. If the correlation between the technological parameters (such as the degree of reduction, the roll diameter, a material's initial thickness, and the friction coefficient) and the model parameters is determined, this two-dimensional FLM approach can enable the efficient calculation of a velocity gradient history evolved across the thickness of a rolled sheet, which is needed for the successful prediction of texture evolution performed by crystal plasticity models.

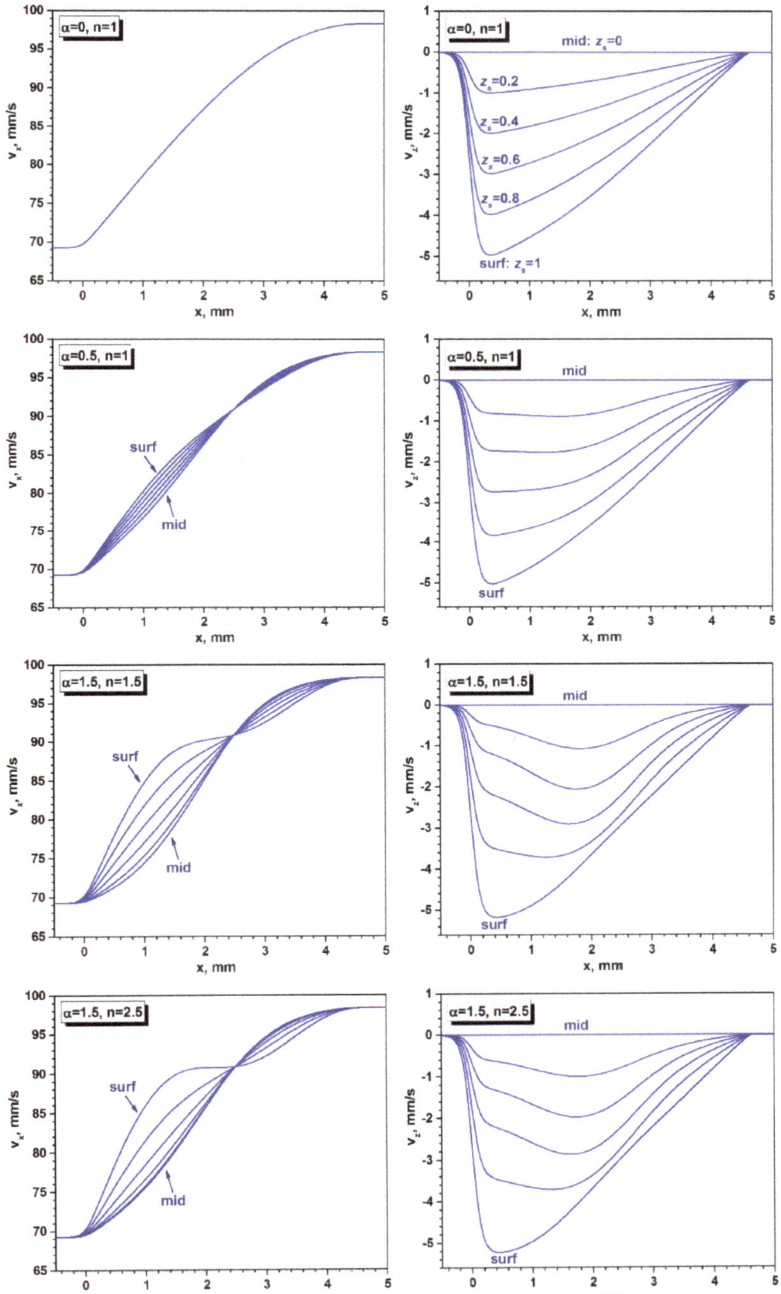

Figure 3. Velocity components (v_x and v_z) calculated by the flow-line modeling (FLM) employed for various model parameters along six flow lines. A sheet of 1.125-mm initial thickness was subjected to a 29.6% reduction with a roll of R = 64.5 mm. The sequence of streamlines was the following: z_s = 1 (surface), z_s = 0.8 (subsurface 1), z_s = 0.6 (subsurface 2), z_s = 0.4 (subsurface 3), z_s = 0.2 (subsurface 4), and z_s = 0 (midplane).

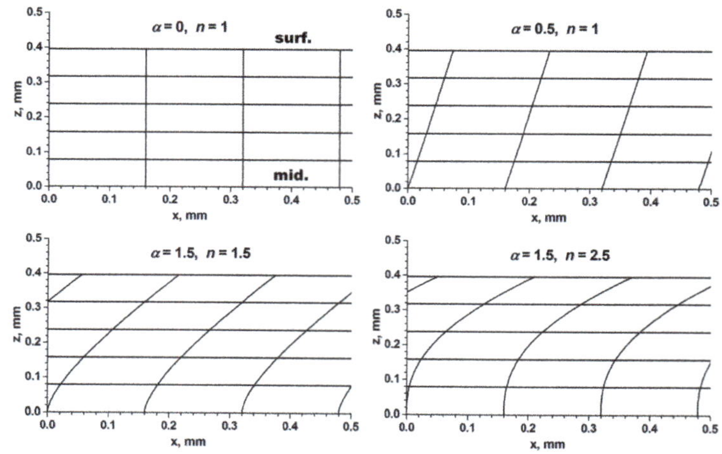

Figure 4. Deformation patterns of the initially rectangular grid (half-thickness) emerged after a 29.6% reduction (initial thickness = 1.125 mm, R = 64.5 mm), as predicted by FLM for various model parameters. The sequence of streamlines was the following: $z_s = 1$ (surface), $z_s = 0.8$ (subsurface 1), $z_s = 0.6$ (subsurface 2), $z_s = 0.4$ (subsurface 3), $z_s = 0.2$ (subsurface 4), and $z_s = 0$ (midplane). In the FLM calculation, the position of a neutral point was identical for all cases, $L_{dN} = 2.5$ mm.

3.2. Finite Element Modeling

In order to define a link between the FLM fitting parameters (α and n) and the processing quantitative indicators, a series of finite element simulations were performed. Since the FLM approach does not account for the anisotropy of properties, the behavior of an isotropic material was studied by FEM. In the first instance, the effect of friction on the deformation flow across the thickness of a rolled Al sheet was studied. The minimum value of the friction coefficient necessary for rolling ($\mu_{\min}$) was approximated based on roll gap geometry, as rolling of a material is not possible without an appropriate friction condition [23].

Equation (8) is

$$\mu_{\min} = \frac{1}{2}\sqrt{\frac{h_f}{R}}\frac{\ln\left(\frac{h_i}{h_f}\right) + \frac{1}{4}\sqrt{\frac{h_i - h_f}{R}}}{\tan^{-1}\sqrt{\frac{h_i}{h_f} - 1}}, \tag{8}$$

For the FEM, rolling trials were simulated with a constant coulomb friction coefficient μ, which exceeded the value of $\mu_{\min}$, in order to avoid both convergence problems and remeshing during the calculation. As Figure 5 reveals, the deformation flow was heterogeneous across the thickness even if the material was isotropic. The deformation patterns that emerged in Figure 5 testify to the fact that distortion of an initially rectangular grid was conditioned by the friction coefficient μ assumed. The distortion of initially orthogonal elements enabled a revealing of the nature of the deformation process across the thickness. Both the surface and subsurface elements (Figure 5a–d) were subjected to more severe shear distortion compared to the midthickness layers. This phenomenon was revealed more explicitly while the distribution of the von Mises strain e_{vM} across the thickness was being analyzed. Since the amount of shear was negligible in the midplane, the strain mode was close to plane strain compression, while both a high value of e_{vM} (Figure 5a–d) and strongly sheared elements in the subsurface region provided evidence for shear localization within the surface layers. It is also obvious in Figure 5a–d that the amount of shear was localized predominantly in the subsurface region and tended to decline in the direction of the midthickness plane independently of friction assumed, whereas increasing the value of μ tended to enhance the level of accumulated strain in the material.

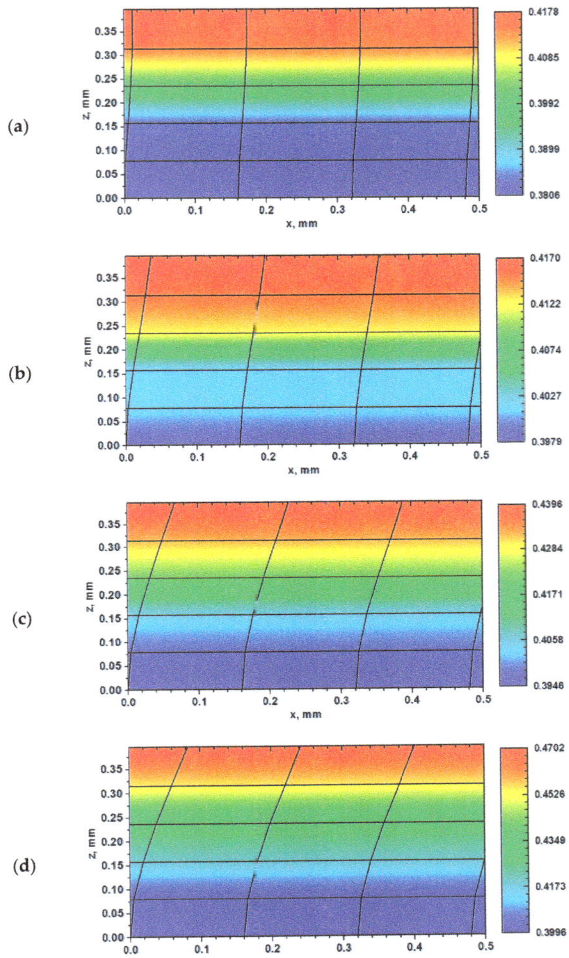

Figure 5. Deformation patterns and von Mises strain distribution across the thickness of a rolled sheet, subjected to a 29.6% thickness reduction (initial thickness = 1.125 mm, R = 64.5 mm), as predicted by finite element modeling for various friction coefficients: (**a**) μ = 0.05; (**b**) μ = 0.1; (**c**) μ = 0.15; (**d**) μ = 0.2. The grid was rectangular prior to deformation.

In total, 53 FEM simulations were performed for various roll gap geometries and diverse friction conditions. These results will be discussed in the following chapters, when both FEM and FLM outputs will be compared. In order to capture the strain gradients in rolling with reasonable accuracy, finite element calculations were performed with a mesh containing 10 elements across the thickness (5 elements for the half-thickness). It is believed that this number of elements is capable of ensuring meaningful simulation precision. Increasing the number of elements might improve accuracy; however, taking into account that a large number of FEM simulations had to be performed within a rational time-bounded frame, the decision was made to work with the fixed number of elements for all roll gap geometries.

3.3. Correlating the Flow-Line Model Parameters with Roll Gap Geometry

A growing body of experimental evidence has clearly shown that TMP parameters have a decisive effect on the deformation behavior of materials [24–27]. This study had a major focus on the cold rolling process, where the flow of a sheet is conditioned by roll gap geometry, straining levels and friction condition. Figure 5 demonstrates the effect of the friction coefficient on the through-thickness strain heterogeneity, as predicted by FEM. Corresponding calculations (Figure 6) were likewise performed for the identical rolling path with FLM [14], where each combination of the α and n parameters was set so as to resemble the deformation patterns of Figure 5. For a meaningful comparison, first, the strain velocity gradient tensor components (L_{ij}) were calculated by both FEM and FLM [14], and afterward the corresponding strains for the time increment Δt were calculated as $\Delta\varepsilon_{ij} = 0.5\ (L_{ij} + L_{ji})\Delta t$. Since in two-dimensional analysis $L_{22} = L_{12} = L_{21} = L_{23} = L_{32} = 0$ (where indices 1, 2, and 3 refer to rolling (RD), transverse (TD), and normal (ND) directions, respectively), the von Mises strain was calculated by means of the total accumulated strain components ε_{11} and $\Delta\varepsilon_{33}$:

$$e_{vM} = \frac{2}{\sqrt{3}}\sqrt{\varepsilon_{11}^2 + \varepsilon_{13}^2}. \tag{9}$$

When comparing Figures 5 and 6, it turns out that FLM was capable of reproducing both the corresponding FEM deformation patterns and the strain distribution across the thickness with reasonable accuracy by employing an appropriate pair of model parameters. Figures 5 and 6 reveal meaningful qualitative agreement, whereas the quantitative differences in the von Mises maps might be explained by simplifications made in the FLM model. The types of von Mises strain distribution patterns shown in Figures 5 and 6 are characteristic of the rolling process, where the high e_{vM} values are attributed to the concentration of shear strain due to friction between the roll and the surface of a sheet.

It should be underlined here that the main advantage of the FLM-type approach is its efficiency, since the deformation history can be simulated within a fraction of a second, which cannot be done with an FEM approach even if the grid is composed of a limited number of elements. The major drawback to the practical implementation of flow-line computations is the determination of α and n, which have to be fitted for each particular case. Therefore, this contribution attempted to correlate the TMP quantitative indicators with the FLM model parameters by employing FEM results. In the FEM simulations, which were performed for a wide range of processing conditions, the coefficient of friction was varied between $\mu_{\min}$ (Equation (8)) and 0.3, which corresponded to wet and dry rolling conditions, respectively. The roll radii in the FEM calculations changed from 64.46 mm to 450 mm, which led to a spectrum of projected contact lengths (L_d) ranging between 3.61 mm and 21.91 mm, whereas the initial thickness of the sheet subjected to cold rolling changed from 1 to 6 mm. These boundary conditions covered the rolling trials performed with both intermediate and large draughts, which accounted for a variation in the contact length-to-mean thickness ratio (L_d/h) from 2.88 to 12.77.

After careful analysis of both the FEM and FLM outputs, the following expressions were developed for the determination of α and n:

$$\alpha = A\psi + B\psi^2 + C\psi^3 + D\psi^4 + E\left(\frac{h_i - h_f}{2L}\right), \tag{10}$$

with

$$\psi = \mu^f\left(\frac{h_i}{R}\right)^g\left(\frac{R}{h_i - h_f}\right)^k\left(\frac{R}{h_f}\right)^l, \tag{11}$$

where $A = 21{,}332$; $B = -2.250 \times 10^2$; $C = -1.841 \times 10^2$; $D = 1.242 \times 10^4$; $E = 4.833 \times 10^{-1}$; $f = 1.833$; $g = 7.892 \times 10^{-1}$; $k = 2.293 \times 10^{-1}$; and $l = 2.283 \times 10^{-1}$. Equation (12) is

$$n = M\mu^r\left(\frac{h_i}{R}\right)^s + N\mu^t\left(\frac{L}{L_{dN}}\right)^u + O\mu^v\left(\frac{h_i + h_f}{R}\right)^w + P\mu^x\left(\frac{L}{h_i + h_f}\right)^y + Q, \tag{12}$$

where $M = 19.369$, $N = 9.463$, $O = -18.606$, $P = -9.086$, $Q = 3.201 \times 10^{-2}$, $r = 9.882 \times 10^{-2}$, $s = 5.120 \times 10^{-3}$, $t = -1.869 \times 10^{-1}$, $u = 6.414 \times 10^{-2}$, $v = -1.045 \times 10^{-1}$, $w = -5.860 \times 10^{-3}$, $x = 2.441 \times 10^{-1}$, and $y = -1.620 \times 10^{-1}$.

(a) (b) (c) (d)

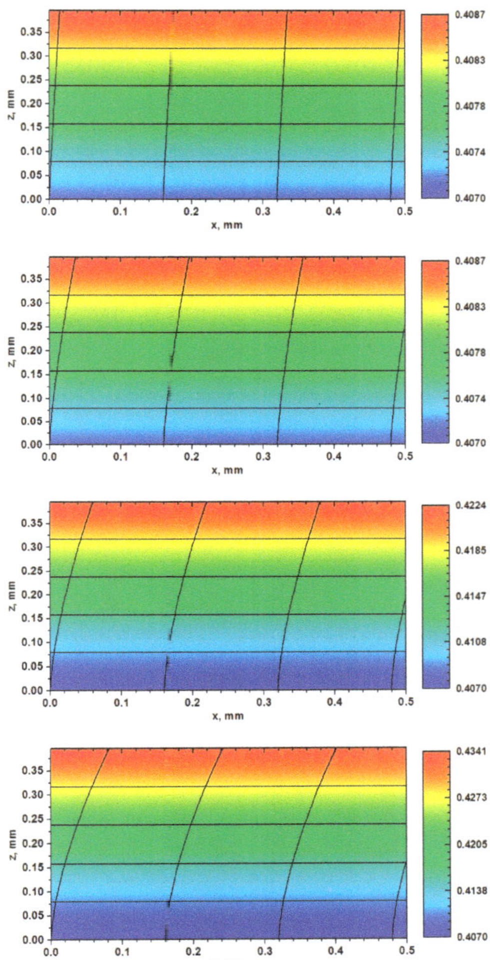

Figure 6. Deformation patterns and von Mises strain distribution across the thickness of a rolled sheet, subjected to a 29.6% thickness reduction (initial thickness = 1.125 mm, R = 64.5 mm), as predicted by the flow-line model employed [14] with the following model parameters: (**a**) $\alpha = 6.12 \times 10^{-2}$, $n = 1.11$; (**b**) $\alpha = 1.55 \times 10^{-1}$, $n = 1.38$; (**c**) $\alpha = 2.69 \times 10^{-1}$, $n = 1.52$; (**d**) $\alpha = 3.84 \times 10^{-1}$, $n = 1.6$. The grid was rectangular prior to deformation. Each set of FLM fitting parameters ensured the best fit for the corresponding finite element modeling (FEM) output (shown in Figure 5a–c).

Equations (10)–(12) clearly indicate that both α and n are functions of roll gap geometry and the friction coefficient μ. Figure 7 reveals the distortion of the initially rectangular grid (only the vertical line is shown for each case) predicted for various roll gap geometries by the FEM and FLM employed [14] (with the model parameters calculated by Equations (10)–(12)). Analyzing the results of the simulations presented in Figure 7, it becomes obvious that both models predicted similar distortion patterns for both extremes of the friction coefficient spectrum. By virtue of assumptions made in the FLM employed, this model was capable of reproducing parabolic-type displacement profiles, while the FEM results revealed more complex displacement shapes for certain boundary conditions (see Figure 7). This was particularly true for low values of both L_d/h and μ. For roll gap geometry with $L_d/h \leq 4$ and friction coefficients below 0.1, the FLM somewhat overestimated the amount of shear on the surface and slightly underestimated the shear contribution within the subsurface layers; however, these inconsistencies tended to vanish when the value of L_d/h became higher than 4.

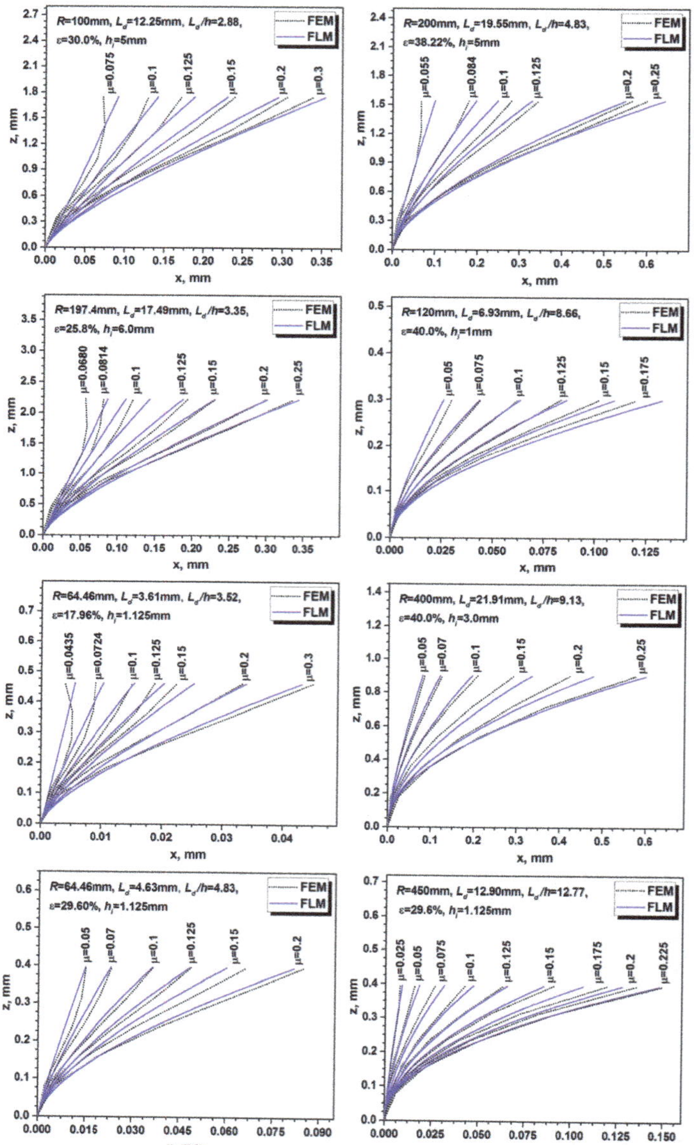

Figure 7. Grid distortions predicted for various roll gap geometries by the FLM employed and FEM. The FLM model parameters were calculated by Equations (10)–(12).

It is clear from Figure 7 that roll gap geometry has a strong impact on the evolution of deformation across the thickness. The effect of rolling conditions on the deformation distribution across the thickness could be analyzed by means of the FLM approach, since the values of the α and n parameters, calculated by Equations (10)–(12), tended to resemble the FEM deformation patterns with reasonable accuracy. In the case of materials (Figure 8a,b) rolled with a thickness reduction of 29.6% and a friction coefficient of 0.1, an increase in the roll radius accounted for more extensive RD–ND displacement, and therefore more severe shear strain evolution evolved in both the surface and subsurface regions.

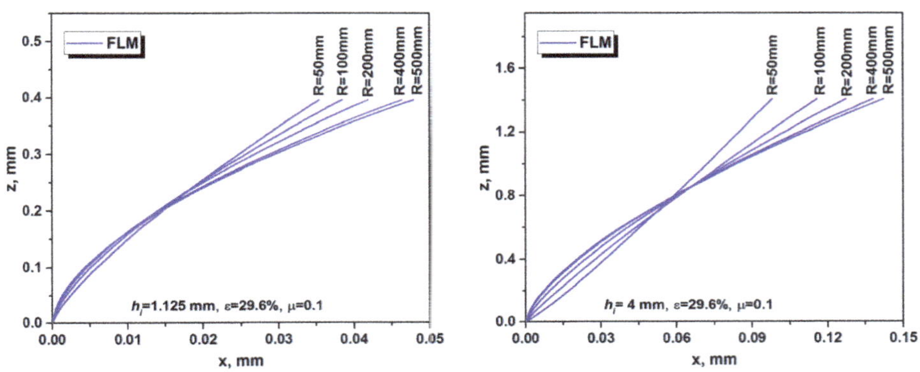

(a) (b)

Figure 8. Deformation patterns predicted for identical rolling reductions and various initial thicknesses by the FLM [14]: (**a**) initial thickness = 1.125 mm; (**b**) initial thickness = 4.0 mm.

It should be underlined that the applicability of the FLM approach [14] is not restricted exclusively to cold rolling; however, the correlation (Equations (10)–(12)) between the FLM model parameters and the process quantitative indicators was limited to an analysis of the cold deformation trials of Al alloys. Defining both α and n by Equations (10)–(12) excluded the effect of temperature, which might be a critical issue while simulating the behavior of materials during hot rolling.

4. Application of Deformation History to Texture Simulation

Al alloys are generally cold-rolled with thickness reductions up to ~85% in several passes; however, in some cases, intermediate annealing can be employed to avoid texture banding, which leads to a roping phenomenon [24,27]. In the current case, the material was subjected to a 29.6% reduction along one path. Preceding the rolling, the annealing process ensured a recrystallization texture in the investigated material. The bulk (overall) texture of the sheet (Figure 9a) measured across the thickness was mainly composed of recrystallization texture components such as Cube ({001}⟨100⟩), Goss ({001}⟨100⟩), and ~ {113}⟨121⟩ orientation ($\varphi_1 = 75°$, $\Phi = 25°$, $\varphi_2 = 45°$). As revealed in Figure 9b, the deformation process induced qualitative texture transformation, where the recrystallization texture components rotated to both α (⟨001⟩//ND) and β-fiber orientations, which is analytically described by the following expression:

$$\left\{\begin{array}{ccc} h, & 1, & h+1 \end{array}\right\}\left\langle \frac{h(h+1)}{3/4-h}, \quad \frac{2h(h+1)}{1/2-h}, \quad \frac{h^2}{h-3/4}+\frac{2h}{h-1/2} \right\rangle, \tag{13}$$

The reference components of the β-fiber calculated by means of Equation (13) enable the tracing of the texture evolution involved in cold rolling. The deformation texture orientations showed an unequal distribution along the β-fiber (Figure 9b). The intensity along this fiber tended to decline with an increase in h (Equation (13)), i.e., the texture intensity followed a downward trend while moving from Copper ($h = 1$) via {213}⟨9 15 11⟩ ($h = 2$) to Brass (h is an infinitely large number). In Figure 9b, a 29.6% rolling reduction did not cause a severe drop in the intensity of the Cube orientation.

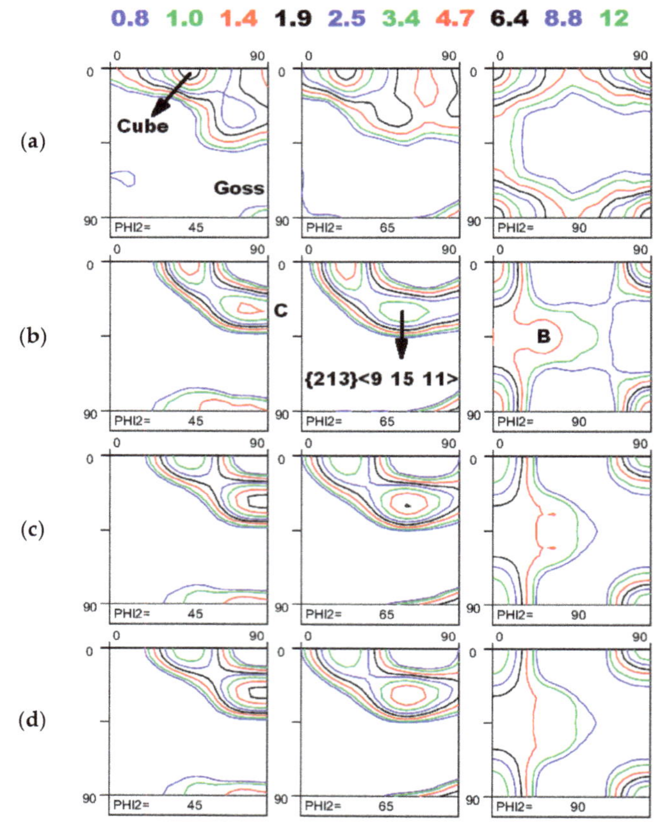

Figure 9. Experimentally measured and simulated overall textures in the investigated material: (**a**) the orientation distribution function (ODF) prior to rolling; (**b**) experimentally measured deformation texture after a 29.6% rolling reduction, where $TI = 2.219$ and $ODF_{max} = 6.03$; (**c**) the ODF simulated with the Alamel model and the strain history calculated by FEM, where $ID = 0.085$, $TI = 2.533$, and $ODF_{max} = 7.57$; (**d**) the ODF simulated with the Alamel model and the strain history predicted by FLM, where $ID = 0.076$, $TI = 2.460$, and $ODF_{max} = 7.30$. The deformation history used for the texture simulations is shown in Figure 10.

Figure 9c,d reveals the evolution of the rolling textures modeled with the Alamel model [5,6] by employing the deformation history in Figure 10. The CP calculations were performed for six equidistant layers (the surface, four subsurface layers, and midthickness plane), and subsequently, the average bulk texture was calculated by merging the individual ODFs. When comparing Figure 9b–d, it becomes apparent that both the experimental and modeled textures followed an identical evolutionary pattern. Similarly to the experimentally measured ODF of Figure 9b, the β-fiber orientations of the modeled counterparts (Figure 9c,d) were well distinguished in the first part of the skeleton line (between {112}⟨111⟩ and {213}⟨9 15 11⟩), while toward the tail of the fiber (running from {213}⟨9 15 11⟩ to {101}⟨121⟩), the individual components tended to vanish.

Inasmuch as the rolling was performed on well-lubricated surfaces on both the sheet and rolling cylinders, the FEM simulation was performed with the μ slightly exceeding the minimum value necessary for rolling ($\mu = 1.5\mu_{min}$), while the corresponding FLM model parameters (α and n) were approximated by Equations (10)–(12). As Figure 9 shows, an approximation of rolling by both the FEM and FLM employed [14] provided a meaningful texture prediction; however, employing various deformation histories accounted for diverse texture development. The quality of the texture simulation was assessed by analyzing qualitative texture indicators (TIDs) such as the ODF maximum value (ODF_{max}), texture index ($TI = \int [f(g)]^2 dg$), and normalized index difference ID. The ID, observed between the experimental $f_{exp}(g)$ and the simulated $f_{sim}(g)$ textures, tended to rise with an increase in diversity between the compared ODFs:

$$ID = \int [f_{exp}(g) - f_{sim}(g)]^2 dg / \int [f_{exp}(g)]^2 dg \tag{14}$$

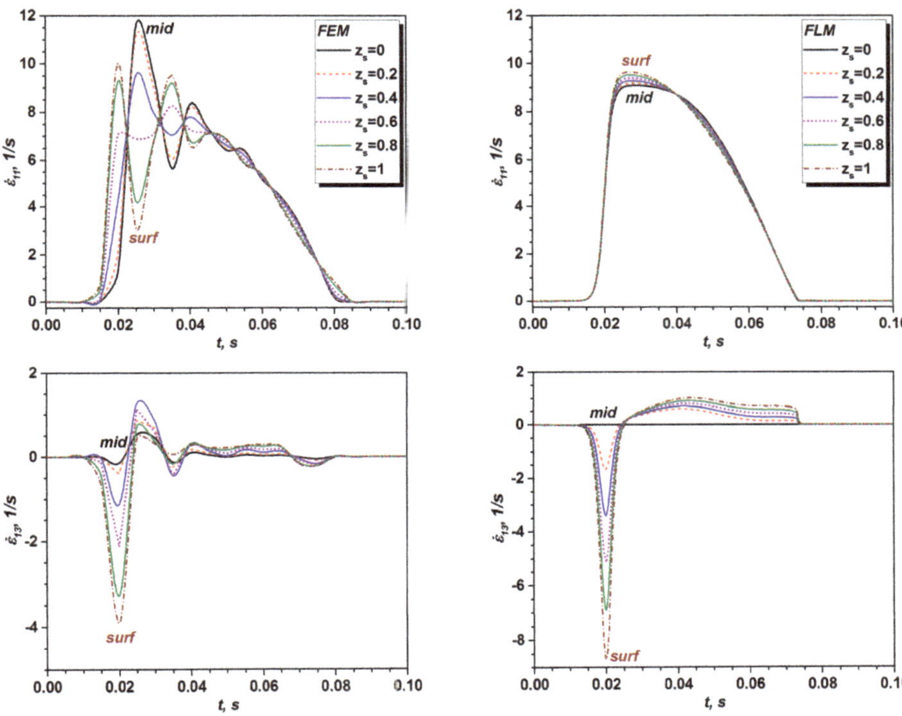

Figure 10. Evolution of strain rate components with time across the thickness of a rolled sheet calculated with both FEM and FLM for a 29.6% reduction (initial thickness = 1.125 mm, R = 64.5 mm). In FEM, the friction coefficient was assumed to be $\mu = 1.5\mu_{min} = 0.053$, while the corresponding FLM fitting parameters were calculated by Equations (10)–(12) ($\alpha = 0.066$, $n = 1.139$).

An examination of the TIDs calculated for Figure 9c,d clearly indicates that the FLM coupled with the CP model was capable of carrying out a texture simulation close to the one performed with the deformation history obtained by means of the finite element model. The employment of both FEM and FLM in the texture simulation led to a slight overestimation of the ODF_{max} value compared to the experimental counterpart, whereas the calculated TIs revealed a close resemblance to the experimentally measured one. Furthermore, the low ID numbers obtained by FEM and FLM (ID(FEM) = 0.085 and ID(FEM) = 0.076) suggest that the FLM employed can be considered to be a potential approach for the computation of strain history in rolling. Although the time necessary for crystal plasticity simulation is hardly affected by the strain mode, the same cannot be said about the calculation of strain history by FEM and FLM, which gives credit to the practical implementation of the latter. The FLM calculation can be performed within a fraction of a second, while FEM requires hours for computation.

The heterogeneity of texture evolution across the thickness of a rolled sheet was accounted for by considering the heterogeneous nature of the strain path evolution in each individual layer (see Figure 10). The through-thickness strain inhomogeneities tended to evolve due to unequal shear distribution. In cold-rolling trials, the sense of balance between the geometric and friction-induced shear components is governed by friction conditions. As seen in Figure 10, first the resulting shear rate $\dot{\varepsilon}_{13}$ was of a negative sense and after that became positive. This implied that initially $\dot{\varepsilon}_{13}$ was tilted toward the roll gap geometry-induced shear components, while later the friction-induced shear tended to prevail [2]. Apart from the quantitative diversity (Figure 10), the major difference between the strain rate patterns calculated by FEM and FLM was that the flow-line model predicted a more gradual change of $\dot{\varepsilon}_{ij}$ components with time, while the corresponding counterparts computed by FEM revealed more oscillating behavior.

Figure 11 reveals both an experimentally measured and simulated evolution of texture in various thickness layers. The ODF of the first layer covered the area between the surface and 3/10 of the half-thickness, the second layer extended from the 3/10 to 3/5 of the half-thickness, and the third layer ranged between 3/5 and the midplane of the sheet. The ODFs of these layers, displayed in Figure 11a, reveal a close qualitative resemblance, while some quantitative diversities can be observed between the individual textures. In order to access the degree of texture heterogeneity (ID_H) between the two layers (i and j), the index difference was expressed as

$$ID_H = \int [f_i(g) - f_j(g)]^2 dg / \int [f^0{}_{\exp}(g)]^2 dg, \tag{15}$$

where $f^0{}_{\exp}(g)$ refers to experimentally measured overall (bulk) texture.

Prior to cold rolling (Figure 11a), the recrystallization process ensured a minimal degree of heterogeneity in the investigated material (see Table 1 for details). The values of ID_H tended to enhance after cold rolling, pointing toward the evolution of texture diversities in various layers ($ID_H(L_{1\text{–}2}) = 6.43 \times 10^{-2}$ and $ID_H(L_{2\text{–}3}) = 2.40 \times 10^{-2}$). This evolutionary pattern was likewise captured by the crystal plasticity calculations, which predicted more homogeneous textures for layers 2 and 3, whereas the extent of heterogeneity rose when the ID_H values for layers 1 and 2 were computed (FLM-CP: $ID_H(L_{1\text{–}2}) = 4.59 \times 10^{-2}$ and $ID_H(L_{2\text{–}3}) = 5.60 \times 10^{-3}$; FEM-CP: $ID_H(L_{1\text{–}2}) = 4.41 \times 10^{-2}$ and $ID_H(L_{2\text{–}3}) = 5.33 \times 10^{-3}$).

Table 1. Heterogeneity quantitative indicators calculated for the textures in Figure 11.

Layers	ID_H (Experiment)		ID_H (FEM)	ID_H (FLM)
	Pre-rolling	After rolling	After rolling	After rolling
Layers 1 and 2	2.94×10^{-3}	6.43×10^{-2}	4.41×10^{-2}	4.59×10^{-2}
Layers 2 and 3	9.09×10^{-3}	2.40×10^{-2}	5.33×10^{-3}	5.60×10^{-3}

In the rolled material (Figure 11b), the intensity of the {001}⟨100⟩ orientation was enhanced from the surface to the midthickness, and this tendency was equally captured by the CP simulations of Figure 11c,d. The same evolutionary trend was observed for the {112}⟨111⟩ component, while the tail of the β-fiber (particularly the {101}⟨121⟩ orientation) tended to weaken from the surface to the middle. This phenomenon was also captured by both the FEM-CP and FLM-CP simulations. The quantitative texture indicators, which allowed for comparisons between the experimentally evolved deformation texture patterns (Figure 11b) and the modeled counterparts (Figure 1c,d), are summarized in Table 2. It is obvious that the results of the simulations showed a close resemblance between the calculated and experimentally measured counterparts. Furthermore, a minimal discrepancy was observed between the FLM- and FEM-based CP predictions.

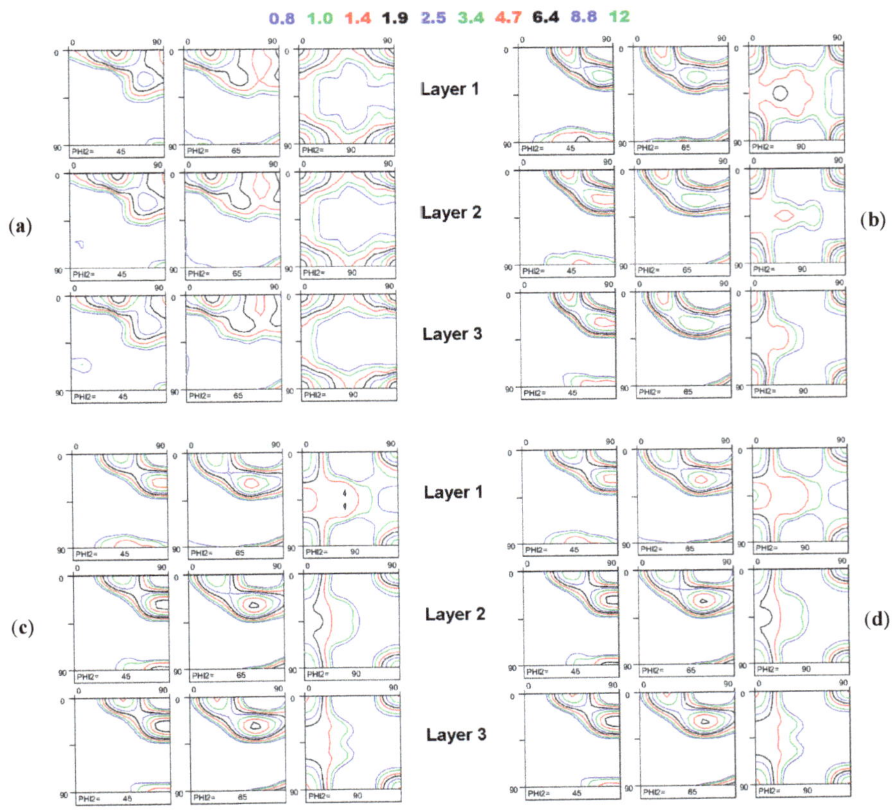

Figure 11. Experimentally observed and simulated texture heterogeneities across the thickness: (**a**) the prerolling texture; (**b**) the measured deformation texture (after a 29.6% reduction); (**c**) the ODF simulated with the Alamel model and the strain path predicted by FEM ($\mu = 0.053$); (**d**) the ODF simulated with the Alamel model and the strain path predicted by the FLM employed ($\alpha = 0.066$, $n = 1.139$).

Table 2. Texture quantitative indicators calculated for the textures in Figure 11.

Layer	FEM			FLM			Experiment	
	ODF_{max}, m.r.d.	*TI*	*ID*	ODF_{max}, m.r.d.	*TI*	*ID*	ODF_{max}, m.r.d.	*TI*
Layer 1	6.0	2.186	0.055	5.73	2.108	0.047	5.99	2.015
Layer 2	8.02	2.645	0.100	7.77	2.571	0.091	6.37	2.353
Layer 3	8.75	2.787	0.089	8.44	2.719	0.083	5.93	3.368

In addition to previously published works [3–11,28], this investigation reveals how the quality of the texture simulation is affected by the choice of strain history approximation. Examining the results of the CP computations shown in Figures 9 and 11, it can be noticed that the Taylor-type homogenization model (Alamel) [5,6] provided both a qualitatively and quantitatively reasonable texture prediction upon the condition that the technique employed for the computation of the strain history was capable of capturing the strain heterogeneity across the thickness. As was shown in recent work [28], the application of other homogenization schemes (VPSC [7] or Cluster V [8]) likewise tends to provide meaningful texture evolutionary patterns with quantitative texture characteristics comparable to the Alamel model. Therefore, the FLM employed [14] can be successfully coupled with an arbitrary CP model.

Although FLM neglected the strain hardening phenomena, while the materials experienced hardening during deformation in the FEM, the result of the CP simulations performed with both FEM and FLM outputs revealed minor differences (Figures 9 and 11). This implies that strain hardening parameters have a slight effect on the quality of texture prediction. Furthermore, approximating the rolling process with plane strain compression or geometric models [10,28] equally provides a reasonable estimate of texture evolution. Even though Taylor-type homogenization models employ strain hardening, varying the values of hardening model parameters for metallic systems, in which the deformation is governed by a single mode (such as a slip), has a negligible effect on the quality of texture prediction. It should be pointed out that this scenario is not applicable to polycrystals, which tend to deform in two diverse modes. In metals deformed by sleep and twinning, the deformation occurs by means of the first mechanism up to a certain strain level, whereas exceeding this straining degree induces another mechanism of deformation (twinning). In this case, the choice of correct hardening parameters is of critical importance. It should be likewise stressed here that when materials experience extensive strain hardening or are subjected to severe plastic deformation, the evolution of local phenomena (such as shear bands) cannot be neglected, since they affect texture development. In contrast to homogenization CP theories, crystal plasticity finite element approaches or CP formulations based on fast-Fourier transforms [4] are capable of predicting macroscopic strain heterogeneities such as through-thickness strain gradients and also providing information on local phenomena (such as the strain/stress evolution on grain boundaries or the development of shear bands). These numerical approaches are computationally demanding and are not discussed in the frame of this study.

In addition to the above-mentioned issues, there are many pros and cons regarding the implementation of analytical models; however, to make the application of FLM approaches more practically attainable, both the upper and lower bounds of implementation should be determined. This investigation predominantly focused on Al alloys with a yield point comparable to 80 MPa; however, even in the frame of a particular series, the yield stress may vary around ~100 MPa. This change in the flow stress will significantly affect the stress state of materials during rolling, which is not accounted for in FLM [14]. In order to determine the applicability limits of the FLM employed, the deformation flow was analyzed at the two ends of the L_d/h spectrum (L_d/h = 2.88 and L_d/h = 12.77) in materials with σ (σ_y = 80 MPa and σ_y = 200 MPa).The results of the FEM simulations presented in Figure 12 suggest that an increase in yield stress from 80 MPa to 200 MPa triggers some variety in the displacement across the thickness at low values of the contact length-to-mean thickness ratio (L_d/h = 2.88), whereas this deviation tends to vanish when rolling is performed with large draughts (L_d/h = 12.77). Analyzing Figures 5–7 and 12, it turns out that the deformation flow in materials, as predicted by both the FEM and FLM simulations, was majorly controlled by roll gap geometry and friction conditions, while the hardening phenomena seemed to have a smaller effect on the corresponding displacement profiles. The inconsistencies shown in Figure 12 did not cause significant qualitative and quantitative texture changes, since minor strain path variations do not affect the evolution of texture drastically (see Figures 9 and 11). This implies that FLM [14] can be employed for texture simulations in Al alloys from diverse series; nevertheless, it should be pointed out that if material properties or processing limits significantly deviate from the above-mentioned bounds, the results should be treated with particular care.

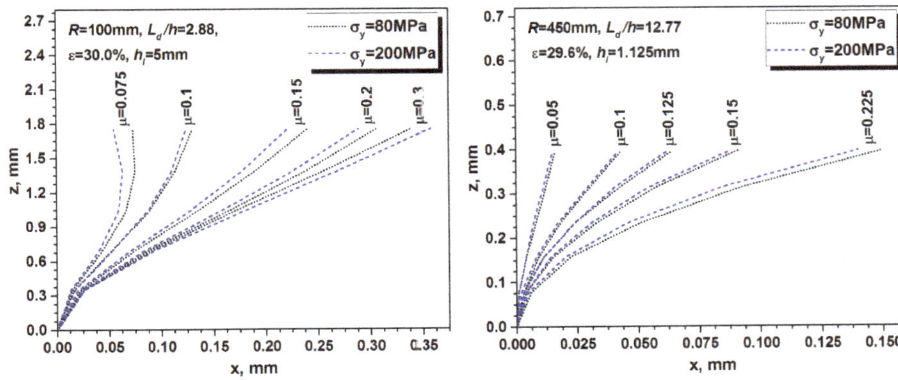

Figure 12. Deformation patterns predicted by FEM for materials with various yield stresses.

5. Conclusions

Careful analysis of both the finite element (FEM) and flow-line (FLM) model outputs led to the conclusion that the FLM model parameters were directly correlated with roll gap geometry and the friction coefficient. Analytical expressions for the determination of α and n were developed in the frame of the contribution, which ensured the implementation of this approach without fitting parameters. The boundary conditions covered the rolling trials with both intermediate and large draughts, which accounted for a variation in the contact length-to-mean thickness ratio between 2.88 and 12.77 and a friction coefficient spectrum changing from 0.025 to 0.3.

The mean-field crystal plasticity (CP) model employed (Alamel), which takes into account short-range grain interaction phenomena, served to produce a through-thickness texture heterogeneity when the strain path was computed either by FEM or FLM. The FLM combined with the CP employed tended to provide a reasonably accurate prediction of bulk texture evolution, and this approach was likewise efficient in capturing the evolution of texture in the individual through-thickness layers. An analysis of quantitative texture indicators clearly demonstrated that the Taylor-type homogenization CP model coupled with the FLM was capable of carrying out a texture simulation close to the one performed with deformation history obtained by means of the finite element model.

The results of the CP modeling point toward the possible practical implementation of FLM approaches, since these computationally efficient methods ensure reasonably accurate texture prediction.

Author Contributions: Conceptualization, J.J.S.; J.J.S; investigation, J.J.S; writing—review and editing, J.J.S; funding acquisition, J.J.S.

Funding: This research was funded by the EFOP-3.6.1-16-2016-00018 project: "improving the role of research + development + innovation in higher education through institutional developments assisting intelligent specialization in Sopron and Szombathely". The work was performed in the frame of the Széchenyi 2020 program: "innovative processing technologies, applications of energy engineering, and implementation of wide-ranging techniques for microstructure investigation".

Conflicts of Interest: The author declares no conflict of interest.

References

1. Raabe, D. *Computational Materials Science*, 1st ed.; WILEY-VCH: New York, NY, USA, 1998; pp. 111–298.
2. Lenard, J.G.; Pietrzyk, M.; Cser, L. *Physical Simulation of The Properties of Hot Rolled Products*, 1st ed.; Elsevier: Oxford, UK, 1999; pp. 61–317.
3. Delannay, L.; Jacques, P.J.; Kalidindi, S.R. Finite element modeling of crystal plasticity with grains shaped as truncated octahedrons. *Int. J. Plast.* **2006**, *22*, 1879–1898. [CrossRef]

4. Lebensohn, R.A.; Kanjarla, A.K.; Eisenlohr, P. An elasto-viscoplastic formulation based on fast Fourier transforms for the prediction of micromechanical fields in polycrystalline materials. *Int. J. Plast.* **2012**, *32–33*, 59–69. [CrossRef]
5. Van Houtte, P.; Li, S.; Seefeldt, M.; Delannay, L. Deformation texture prediction: From the Taylor model to the advanced Lamel model. *Int. J. Plast.* **2005**, *21*, 589–624. [CrossRef]
6. Van Houtte, P.; Kanjarla, A.K.; Van Bael, A.; Seefeldt, M.; Delannay, L. Multiscale modeling of the plastic anisotropy and deformation texture of polycrystalline materials. *Eur. J. Mech. A/Solids* **2006**, *25*, 634–648. [CrossRef]
7. Lebensohn, R.A.; Tome, C.N. A self-consistent anisotropic approach for the simulation of plastic deformation and texture development of polycrystals: Application to zirconium alloys. *Acta Metall. Mater.* **1993**, *41*, 2611–2624. [CrossRef]
8. Xie, Q.; Van Bael, A.; Sidor, J.; Moerman, J.; Van Houtte, P. A new cluster type model for the simulation of textures of polycrystalline metals. *Acta Mater.* **2014**, *69*, 175–186. [CrossRef]
9. Mao, W. On the Taylor principles for plastic deformation of polycrystalline metals. *Front. Mater. Sci.* **2016**, *10*, 335–345. [CrossRef]
10. Engler, O.; Huh, M.-Y.; Tome, C.N. A study of through-thickness texture gradients in rolled sheets. *Metall. Mater. Trans. A* **2000**, *31*, 2299–2315. [CrossRef]
11. Sidor, J.J.; Petrov, R.H.; Xie, Q.; Van Houtte, P.; Kestens, L.A.I. Evaluation of crystallographic changes and plastic strain ratio in Al alloys. *Mater. Sci. Technol.* **2017**, *33*, 667–677. [CrossRef]
12. Zhang, S.H.; Zhao, D.W.; Gao, C. The calculation of roll torque and roll separating force for broadside rolling by stream function method. *Int. J. Mech. Sci.* **2012**, *57*, 74–78. [CrossRef]
13. Dogruoglu, A.N. On constructing kinematically admissible velocity fields in cold sheet rolling. *J. Mater. Process. Technol.* **2001**, *110*, 287–299. [CrossRef]
14. Decroos, K.; Sidor, J.; Seefeldt, M. A new analytical approach for the velocity field in rolling processes and its application in through-thickness texture prediction. *Metall. Mater. Trans. A* **2014**, *45*, 948–961. [CrossRef]
15. Beausir, B.; Toth, L.S. *Microstructure and Texture in Steels and Other Materials*, 1st ed.; Haldar, A., Suwas, S., Bhattacharjee, D., Eds.; Springer: London, UK, 2009; pp. 415–417.
16. Tian, Y.; Guo, Y.-H.; Wang, Z.-D.; Wang, G.-D. Analysis of Rolling Pressure in Asymmetrical Rolling Process by Slab Method. *J. Iron Steel Res. Int.* **2009**, *16*, 22–26. [CrossRef]
17. Aboutorabi, A.; Assempour, A.; Afrasiab, H. Analytical approach for calculating the sheet output curvature in asymmetrical rolling: In the case of roll axis displacement as a new asymmetry factor. *Int. J. Mech. Sci.* **2016**, *105*, 11–22. [CrossRef]
18. Liu, Y.-M.; Ma, G.-S.; Zhang, D.-H.; Zhao, D.-W. Upper bound analysis of rolling force and dog-bone shape via sine function model in vertical rolling. *J. Mater. Process. Technol.* **2015**, *223*, 91–97. [CrossRef]
19. Rajmohan, N.; Szpunar, J.A. A new model for recrystallization of heavily cold-rolled aluminum using orientation-dependent stored energy. *Acta Mater.* **2000**, *48*, 3327–3340. [CrossRef]
20. Sidor, J.J.; Decroos, K.; Petrov, R.H.; Kestens, L.A.I. Evolution of recrystallization textures in particle containing Al alloys after various rolling reductions: Experimental study and modeling. *Int. J. Plast.* **2015**, *66*, 119–137. [CrossRef]
21. Athreya, C.N.; Mukilventhan, A.; Suwas, S.; Vedantam, S.; Sarma, V.S. Influence of the mode of deformation on recrystallisation behaviour of titanium through experiments, mean field theory and phase field model. *Model. Simul. Mater. Sci. Eng.* **2018**, *26*, 035004. [CrossRef]
22. van Houtte, P. *MTM-FHM Software*, 2nd ed.; MTM-KU Leuven: Leuven, Belgium, 1995; pp. 1–20.
23. Avitzur, B. Friction-aided strip rolling with unlimited reduction. *Int. J. Mach. Tool. Des. Res.* **1980**, *20*, 197–210. [CrossRef]
24. Humphreys, J.; Rohrer, G.S.; Rollett, A. *Recrystallization and Related Annealing Phenomena*, 3rd ed.; Elsevier: Amsterdam, The Netherlands, 2017; pp. 14–135.
25. Schneider, M.; Werner, F.; Langenkämper, D.; Reinhart, C.; Laplanche, G. Effect of Temperature and Texture on Hall–Petch Strengthening by Grain and Annealing Twin Boundaries in the MnFeNi Medium-Entropy Alloy. *Metals* **2019**, *9*, 84. [CrossRef]
26. Lei, B.; Chen, G.; Liu, K.; Wang, X.; Jiang, X.; Pan, J.; Shi, Q. Constitutive Analysis on High-Temperature Flow Behavior of 3Cr-1Si-1Ni Ultra-High Strength Steel for Modeling of Flow Stress. *Metals* **2019**, *9*, 42. [CrossRef]

27. Bennett, T.; Sidor, J.; Petrov, R.H.; Kestens, L.A.I. The effect of intermediate annealing on texture banding in aluminum alloy 6016. *Adv. Eng. Mater.* **2010**, *12*, 1018–1023. [CrossRef]
28. Sidor, J.J. Deformation texture simulation in Al alloys: Continuum mechanics and crystal plasticity aspects. *Model. Simul. Mater. Sci. Eng.* **2018**, *26*, 085011. [CrossRef]

metals

MDPI

Article

Numerical Investigation of Secondary Deformation Mechanisms on Plastic Deformation of AZ31 Magnesium Alloy Using Viscoplastic Self-Consistent Model

Yong Lian [1], Li Hu [2,*], Tao Zhou [2], Mingbo Yang [2] and Jin Zhang [1]

[1] Institute for Advanced Materials and Technology, University of Science and Technology Beijing, Beijing 100083, China; liany09@126.com (Y.L.); zhangjin@ustb.edu.cn (J.Z.)

[2] College of Material Science and Engineering, Chongqing University of Technology, Chongqing 400054, China; zt19811118@cqut.edu.cn (T.Z.); yangmingbo@cqut.edu.cn (M.Y.)

* Correspondence: huli@cqut.edu.cn; Tel.: +86-173-5842-8920

Received: 2 December 2018; Accepted: 28 December 2018; Published: 5 January 2019

Abstract: Uniaxial tension and compression of AZ31 magnesium alloy were numerically investigated via the viscoplastic self-consistent (VPSC) model to shed a light on the effect of secondary deformation mechanisms (prismatic <a> slip, pyramidal <c+a> slip, and $\{10\bar{1}1\}$ contraction twinning) during plastic deformation. The method adopted in the present study used different combinations of deformation mechanisms in the VPSC modeling. In terms of the pyramidal <c+a> slip, it served as the first candidate for sustaining the extra plastic strain during the plastic deformation. The improvement of activity in the pyramidal <c+a> slip contributed to the increase in the mechanical response and the splitting of pole densities in $\{0002\}$ pole figure during uniaxial tension. As for the prismatic <a> slip, its increasing activity was not only conducive to the improvement of flow stress in mechanical response, but also responsible for the splitting of pole densities in $\{0002\}$ pole figure during uniaxial compression. With respect to the $\{10\bar{1}1\}$ contraction twinning, it had a negligible influence on the plastic deformation of AZ31 magnesium alloy in terms of the mechanical response as well as the slip and the twinning activities. However, it is better to include the $\{10\bar{1}1\}$ contraction twinning in the VPSC modeling to more accurately predict the texture evolution.

Keywords: magnesium alloy; deformation mechanisms; plastic deformation; polycrystal plasticity modeling

1. Introduction

The exploitation of magnesium and its alloys for an increasing diversity of applications in the fields of materials science and industry is derived from their remarkable attractiveness including low density, high specific stiffness and strength, good damping property, etc. [1,2]. These characteristics make magnesium and its alloys one of the light-weight materials with a huge potential in tackling the current energy crisis [3]. Among all magnesium alloys, the AZ31 magnesium alloy would be one of the most widely applied. However, owing to its anisotropy in mechanical response and poor formability at room temperature, many challenges with respect to metal forming and structural application still remain to be solved. The anisotropic plastic deformation and poor formability of the AZ31 magnesium alloy at room temperature resulted from the activated deformation mechanisms during the plastic deformation. In general, it is accepted that the ductility of metallic materials is mainly associated with the number of activated slip/twinning systems during the plastic deformation. As for the AZ31 magnesium alloy with the hexagonal close-packed (HCP) crystal structure, the known slip/twinning systems include $\{0001\} < 11\bar{2}0 >$ (basal <a>) slip, $\{10\bar{1}0\} < 11\bar{2}0 >$ (prismatic <a>) slip, $\{11\bar{2}2\} < 11\bar{2}3 >$

(pyramidal <c+a>) slip, $\{10\bar{1}2\}$ extension twinning, and $\{10\bar{1}1\}$ contraction twinning [4]. Among these deformation mechanisms, the basal <a> slip and the $\{10\bar{1}2\}$ extension twinning had been confirmed to be the main deformation mechanisms at room temperature to accommodate the plastic strain along and perpendicular to the crystal c-axis, respectively, as they possess a relatively lower critical resolved shear stress (CRSS) in comparison with the other deformation mechanisms [4,5]. However, the remaining deformation mechanisms including the non-basal slip (prismatic <a> slip and pyramidal <c+a>) slip) as well as the $\{10\bar{1}1\}$ contraction twinning also play important roles in the plastic deformation of AZ31 magnesium alloy at room temperature as determined through the experimental and modeling investigation [6,7]. In the present study, due to their contributions to plastic deformation, these three deformation mechanisms are termed as secondary deformation mechanisms. In addition, our literature search has shown that there seems to be a lack of study focusing on the roles of these secondary deformation mechanisms during the plastic deformation of AZ31 magnesium alloy.

To date, there is still a lack of efficient and convenient experimental observation techniques in revealing the plastic deformation mechanisms of metallic materials. Consequently, the numerical investigation on the basis of crystal plasticity theory has become a perfect candidate for clarifying the influence of deformation mechanisms during plastic deformation. As for the AZ31 magnesium alloy, researchers have developed numerous models in consideration of the dominant deformation mechanisms to predict the anisotropic mechanical response and texture evolution [8–13]. Among these models, the viscoplastic self-consistent (VPSC) model can be termed as one of the most adopted models due to its efficiency and accuracy in modeling the plastic deformation of metallic materials. Therefore, in the present study, the VPSC model is used to shed a light on the roles of these secondary deformation mechanisms of AZ31 magnesium alloy during the plastic deformation, which could lay foundations for extending its application at room temperature. The method adopted in the present study is to systematically change the embedded deformation mechanisms during the VPSC modeling; however, the input data in terms of initial orientation and boundary conditions remain unchanged.

2. VPSC-Based Crystal Plasticity Modeling

In the present study, the viscoplastic self-consistent (VPSC) model developed by Lebensohn et al. [14] is utilized to predict the macroscopic stress–strain response, slip and twinning activities, and texture evolution of the AZ31 magnesium alloy during uniaxial tension and compression. The core of VPSC model is that it can account for the full anisotropy of properties and responses of the single crystals by treating the constituting individual grains as anisotropic viscoplastic ellipsoidal inclusions [15] inside a homogeneous effective medium (HEM), which represents the average constitutive behavior of the polycrystalline aggregate. The interaction between each grain and HEM can be described by different linearization assumptions embedded in the VPSC model [16]. To retain self-consistency and to ensure strain compatibility as well as stress equilibrium during the numerical modeling, the condition that the average strain rate of individual grains has to be consistent with the applied macroscopic magnitude is enforced in the VPSC model [17]. As the detailed procedure for calculation in the VPSC model has been well documented by Li et al. [18], only the key equations about the hardening model and twinning model are briefly presented here.

The VPSC model is based on the physical shear mechanisms of slip and twinning during the plastic deformation. The activation of the physical shear mechanisms is mainly associated with the corresponding CRSS, whose evolution is totally described by the hardening model. To date, several hardening models have been successfully developed and applied in the VPSC model including a modified Voce model, the mechanical threshold stress (MTS) model, and some dislocation density-based models [18–20]. In the present study, though it is in fact empirical, a modified Voce model is applied because of its simplification during parameter calibration and its accuracy in terms of predicting the mechanical response and the texture evolution during the plastic deformation [21,22].

The modified Voce model relates the evolution of CRSS with the accumulative shear strain of all the activated deformation mechanisms including the slip and the twinning systems.

$$\tau_c^\alpha = \tau_0^\alpha + (\tau_1^\alpha + \theta_1^\alpha \Gamma)(1 - exp(-\frac{\theta_0^\alpha}{\tau_1^\alpha}\Gamma) \tag{1}$$

where τ_c^α represents the instantaneous CRSS of each slip/twinning system α. The expressions τ_0^α and θ_0^α stand for initial CRSS and initial hardening rate, respectively, whereas τ_1^α is associated with the back-extrapolated CRSS and θ_1^α refers to the asymptotic hardening rate. The expression Γ stands for the total accumulated shear strain in all slip and twinning systems.

It is generally accepted that there exist interactions among the slip and the twinning systems, hence the propagation in a slip or a twinning system would be impeded by other slip and twinning systems. This phenomenon has been taken into consideration in the VPSC model by introducing the coupling coefficient $h^{\alpha\alpha'}$.

$$\Delta\tau_c^\alpha = \frac{d\tau_c^\alpha}{d\Gamma}\sum h^{\alpha\alpha'}\Delta\gamma^{\alpha'} \tag{2}$$

where $\Delta\tau_c^\alpha$ refers to the increment of CRSS in a slip or a twinning system, and it mainly depends on the increment of shear strain in each slip and twinning systems. The expression $h^{\alpha\alpha'}$ in the case of $(\alpha = \alpha')$ represents the self-hardening coefficient, whereas $h^{\alpha\alpha'}$ in the case of $(\alpha \neq \alpha')$ is known as the latent-hardening coefficient.

The embedding of twinning in modeling the plastic deformation of metallic materials has been successfully obtained by applying the predominant twin reorientation (PTR) scheme proposed by Tomé et al. [23]. In the PTR approach, both the shear strain γ^α and the corresponding twin volume fraction $f^\alpha = \gamma^\alpha/\gamma^{tw}$ associated with each twin system α within a single grain are tracked at each incremental step (γ^{tw} refers to the characteristic twinning shear). Since treating each twinned fraction as a newly generated grain is not numerically feasible in the PTR scheme a given grain is entirely reoriented only if the predominant twinning system exceeds a threshold value V^{thres} as follows.

$$V^{thres} = A^{th1} + A^{th2}\frac{V^{eff}}{V^{accum}} \tag{3}$$

where A^{th1} and A^{th2} are the two parameters associated with the evolution of twin volume fraction during the plastic deformation, and they could have different values with respect to various twinning modes. In the present study, for the $\{10\bar{1}2\}$ extension twinning, A^{th1}= 0.7 and A^{th2}= 0.1, whereas for the $\{10\bar{1}1\}$ contraction twinning, A^{th1}= 0 05 and A^{th2}= 0.05 [24]. Here, V^{eff} refers to the volume fraction of the twin-terminated grains and V^{accum} represents the volume fraction of the twinned regions.

3. Results and Discussion

In the present study, a hot-rolled AZ31 magnesium alloy plate with a strong basal texture is considered as shown in Figure 1, where the initial {0002} pole figure demonstrates that the c-axis of most grains in the sheet is close to the normal direction (ND). The specific contents about fabricating this AZ31 magnesium alloy sheet and the procedure of measuring its initial texture have been well documented in the literature [24]; hence they are not discussed here. The input data about crystallographic orientations that are needed in the VPSC modeling are then extracted from the measured texture. By using the method of discretizing the orientation distribution function (ODF) via the toolbox MTEX in MATLAB code, which is depicted in the literature [25], 1600 discrete orientations with equal volume fractions are finally used for the VPSC modeling. As for the boundary conditions, which is also key input information in the VPSC modeling, a combination of velocity gradient components and stress components are imposed in the present study to predict the uniaxial tension and compression of the AZ31 magnesium alloy in different loading paths. This operation allows for the ovalization of the deformed sample [26]. In all simulations, a true strain of 0.15

is enforced and a strain rate 0.001 s^{-1} is applied in each loading direction, which is equal to the macroscopic strain rate in uniaxial tension and compression experiments as depicted in the literature [24]. The mainly reported deformation mechanisms involving the plastic deformation of AZ31 magnesium alloy at room temperature are $\{0001\} < 11\bar{2}0 >$ (basal <a>) slip, $\{10\bar{1}0\} < 11\bar{2}0 >$ (prismatic <a>) slip, $\{11\bar{2}2\} < 11\bar{2}3 >$ (pyramidal <c+a>) slip, $\{10\bar{1}2\}$ extension twinning, and $\{10\bar{1}1\}$ contraction twinning. These deformation mechanisms are considered in VPSC simulations. Moreover, $\{10\bar{1}1\} - \{10\bar{1}2\}$ double twinning is also allowed in the present study. In addition, the affine linearization scheme is applied in the present study to relate the responses of single grains to the response of the polycrystal aggregate. Wang et al. [16] have confirmed that the affine scheme gives the best accordance between the experimental results and the corresponding simulated ones for the AZ31 magnesium alloy. The exponent of rate sensitivity $n = 20$ is applied in the present study, which is a recommended value in the VPSC manual to simulate an almost rate-insensitive deformation.

Figure 1 shows the mechanical responses during through-thickness compression (TTC) along the normal direction (ND), in-plane compression (IPC) along the rolling direction (RD), and in-plane tension (IPT) along the transverse direction (TD). These experimental data are from the work of Chapuis et al. [24]. Using the corresponding material parameters identified by Chapuis et al. [24], as shown in Table 1, the predicted stress–strain curves are obtained and are also shown in Figure 1. Though during IPC along RD, the flow stresses between the true strain of 0.04 and 0.07 exhibit some discrepancy, the consistency between the global mechanical responses and the corresponding simulated ones confirms the validity of the adopted materials.

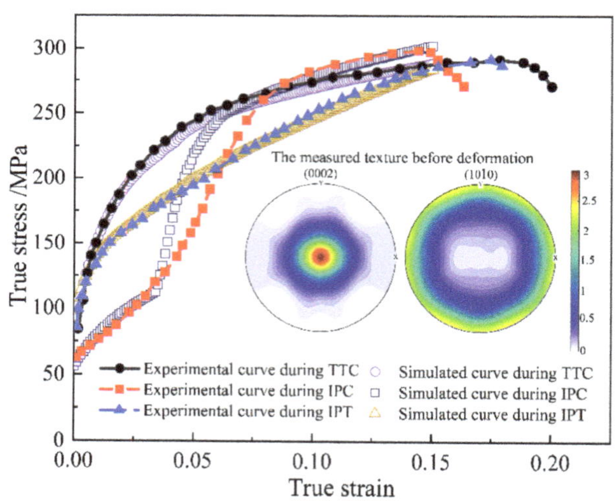

Figure 1. The comparison between the experimental stress–strain curves (lines and symbols) and the corresponding simulated stress–strain curves (symbols) of the AZ31 magnesium alloy under the uniaxial compression along the normal direction (ND), the uniaxial compression along the rolling direction (RD), and the uniaxial tension along the transverse direction (TD) at the true strain of 0.15. (The initial texture and experimental mechanical responses are from Chapuis et al. [24]). TTC, through-thickness compression; IPT, in-plane tension; and IPC, in-plane compression.

Table 1. Material parameters of AZ31 magnesium alloy used in viscoplastic self-consistent (VPSC) modeling.

Mode	τ_0	τ_1	θ_0	θ_1	$h^{\alpha\alpha'}$
Basal <a>	5 MPa	25 MPa	300 MPa	60 MPa	1.0; 1.0; 1.0; 3.0; 3.0
Prismatic <a>	60 MPa	25 MPa	2000 MPa	170 MPa	1.0; 1.0; 1.0; 1.1; 1.1
Pyramidal <c+a>	70 MPa	105 MPa	3500 MPa	60 MPa	1.0; 1.0; 1.0; 1.1; 1.1
$\{10\bar{1}2\}$	30 MPa	0 MPa	0 MPa	0 MPa	1.0; 1.0; 1.0; 1.0; 1.0
$\{10\bar{1}1\}$	200 MPa	30 MPa	1000 MPa	0 MPa	0.0; 1.0; 0.0; 1.0; 0.0

Many works conducted by different researchers have verified that the basal <a> slip and the $\{10\bar{1}2\}$ extension twinning are the main deformation mechanisms with the relatively lower CRSS values at room temperature, whereas the prismatic <a> slip, the pyramidal <c+a> slip, and the $\{10\bar{1}1\}$ contraction twinning serve as secondary deformation mechanisms with the relatively higher CRSS values [5,27,28]. The focus of the present study is to numerically investigate the effects of these secondary deformation mechanisms during the plastic deformation of AZ31 magnesium alloy, hence four different combinations of deformation mechanisms are constructed and then applied in all subsequent VPSC simulations, as shown in Table 2. The first combination of deformation mechanisms is termed as combination A and it consists of basal <a> slip, prismatic <a> slip, pyramidal <c+a> slip, $\{10\bar{1}2\}$ extension twinning, and $\{10\bar{1}1\}$ contraction twinning. It serves as a benchmark in VPSC modeling. The second combination of deformation mechanisms is termed as combination B, which contains basal <a> slip, pyramidal <c+a> slip, $\{10\bar{1}2\}$ extension twinning, and $\{10\bar{1}1\}$ contraction twinning. The third combination of deformation mechanisms is termed as combination C, which includes basal <a> slip, prismatic <a> slip, $\{10\bar{1}2\}$ extension twinning, and $\{10\bar{1}1\}$ contraction twinning. The fourth combination of deformation mechanisms is termed as combination D and it is made up of basal <a> slip, prismatic <a> slip, pyramidal <c+a> slip, and $\{10\bar{1}2\}$ extension twinning.

Table 2. Combinations of deformation mechanisms used in VPSC modeling. (Symbol ● indicates the involved deformation mechanism.)

Designation	Basal <a>	Prismatic <a>	Pyramidal <c+a>	$\{10\bar{1}2\}$	$\{10\bar{1}1\}$
Combination A	●	●	●	●	●
Combination B	●		●	●	●
Combination C	●	●		●	●
Combination D	●	●	●	●	

Figure 2 demonstrates the predicted stress–strain curves during IPC along RD and the corresponding slip and twinning activities in the cases of different combinations of deformation mechanisms. As shown in Figure 2a, there exists little discrepancy between the stress–strain curve in the case of combination A and the corresponding one in the case of combination D. On the contrary, both the stress–strain curves in the cases of combination B and combination C are higher than the corresponding one in the case of combination A. It can be seen from Figure 2b that when the prismatic <a> slip is eliminated from the adopted deformation mechanisms, there exist almost no changes with respect to the activities in the $\{10\bar{1}2\}$ extension twinning and the $\{10\bar{1}1\}$ contraction twinning, and only a minor change appears in the activity of basal <a> slip. However, the evolution of activity in the pyramidal <c+a> slip changes obviously and the increasing activity of the pyramidal <c+a> slip in the case of combination B is nearly equal to the total activity of the prismatic <a> slip in the case of combination A. In addition, it is important to note that the actual activity of the pyramidal <c+a> slip in the case of combination A and combination B are quite close to each other at the true stain of 0.15. As for the pyramidal <c+a> slip, when it is excluded during IPC along RD, as shown in Figure 2c, the evolution laws in the basal <a> slip and the prismatic <a> slip as well as in the $\{10\bar{1}2\}$ extension twinning and the $\{10\bar{1}1\}$ contraction twinning change obviously. Moreover, the $\{10\bar{1}1\}$ contraction twinning possesses the biggest increase in the case of combination C. With respect to the $\{10\bar{1}1\}$ contraction twinning, when it is not embedded in VPSC modeling, there seems to be little influence on the activities of other deformation mechanisms as shown in Figure 2d. Based on the aforementioned analysis with respect to the slip and twinning activities in VPSC modeling with different combinations of deformation mechanisms, it can be noted that the aforementioned discrepancy with respect to the stress–strain curves during IPC along RD is actually derived from the various roles of each deformation mechanism during the plastic deformation of AZ31 magnesium alloy, which has been reported by Chapuis et al. [24].

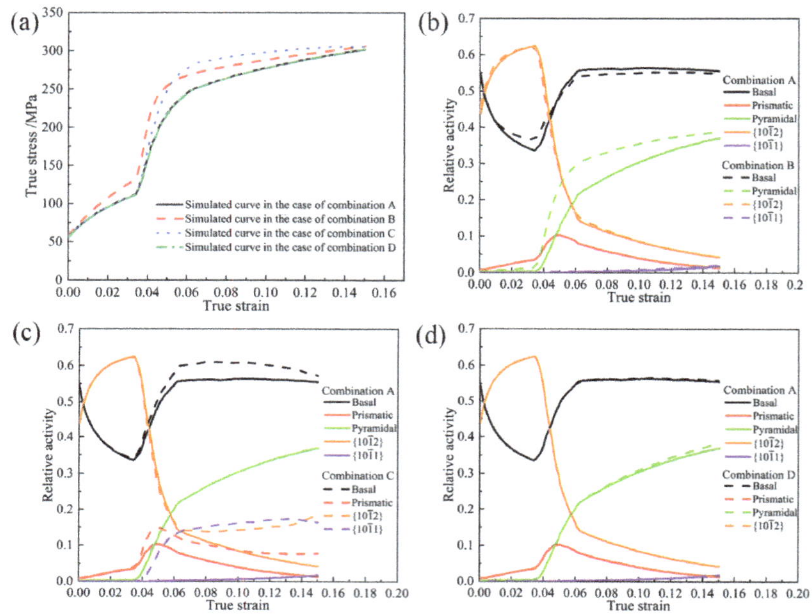

Figure 2. (**a**) The predicted stress–strain curves of AZ31 magnesium alloy under IPC along RD; (**b**–**d**) the corresponding slip and twinning activities predicted by the VPSC modeling in the cases of combination B, combination C, and combination D, respectively.

Figure 3 demonstrates the mechanical responses as well as the slip and twinning activities during IPT along TD. Similar to the evolution laws of stress–strain curves as shown in Figure 2a, the mechanical responses in Figure 3a possess larger flow stresses in the cases of combination B and combination C than in the cases of combination A and combination D. Moreover, in the case of combination B, the growth with respect to the flow stress is bigger than the corresponding one in the case of combination C. The difference in mechanical responses can also be related to the slip and twinning activities. As shown in Figure 3b, in the case of combination A, the basal <a> slip and the prismatic <a> slip possess the dominant activities during IPT along TD. When the prismatic <a> slip is not considered in the VPSC modeling, namely in the case of combination B, the increment in the activity of pyramidal <c+a> slip is the highest in comparison with the corresponding increment with respect to the basal <a> slip. As for the pyramidal <c+a> slip, when it is not embedded in the adopted deformation mechanisms during the VPSC modeling, it only results in a slight increase of activity in the prismatic <a> slip, as shown in Figure 3c. With respect to the $\{10\bar{1}1\}$ contraction twinning, Figure 3d illustrates that whether it exists or not, the activities in other deformation mechanisms would not be influenced during IPT along TD. Hence, it has a negligible influence on the mechanical response of AZ31 magnesium alloy during IPT along TD.

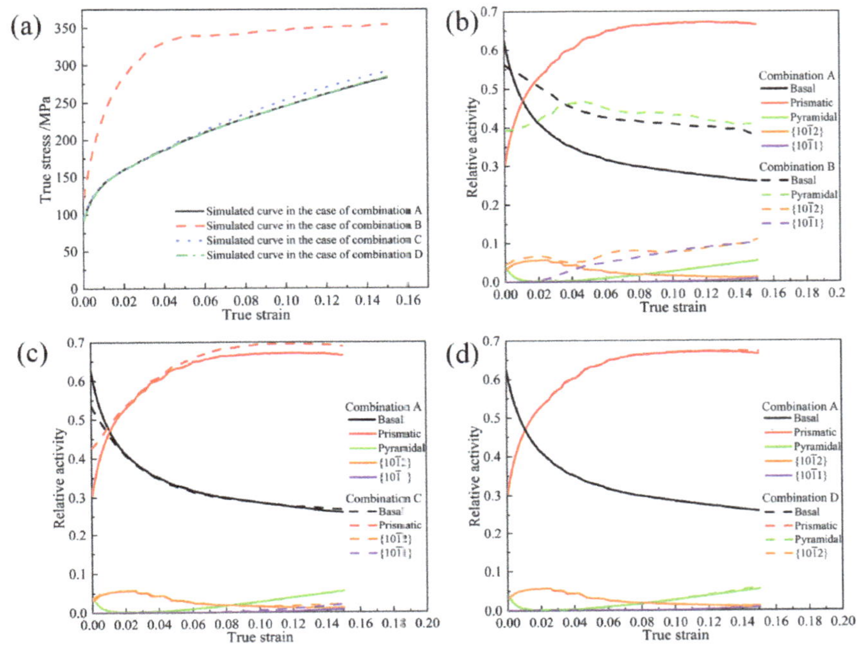

Figure 3. (**a**) The predicted stress–strain curves of AZ31 magnesium alloy under IPT along TD; (**b**–**d**) the corresponding slip and twinning activities predicted by VPSC modeling in the cases of combination B, combination C, and combination D, respectively.

It is generally accepted that the slip and twinning activities have a direct influence on the texture evolution of AZ31 magnesium alloy during the plastic deformation; hence, the predicted textures during IPC along RD and IPT along TD are also analyzed in the present study using VPSC modeling with different combinations of deformation mechanisms, as shown in Figure 4. As for the IPC along RD, Figure 4a,c,g demonstrate that the distribution of pole densities in $\{0002\}$ and $\{10\bar{1}0\}$ pole figures are nearly the same in the cases of combination A, combination B, and combination D. However, in the case of combination C, the splitting of pole densities is obvious around RD, as shown in Figure 4e, and there also exist some pole densities close to ND. In contrast, during IPT along TD, the $\{0002\}$ and $\{10\bar{1}0\}$ pole figures exhibit only minor changes in the case of combination A, combination C, and combination D, as shown in Figure 4b,f,h, respectively. However, in the case of combination B, there exists an obvious splitting of pole densities around ND in the $\{0002\}$ pole figure.

The aforementioned difference with respect to the predicted texture can also be related to the slip and twinning activities during the plastic deformation. Based on the aforementioned analysis on the slip and twinning activities during IPC along RD and IPT along TD, an agreement can be made that the pyramidal <c+a> slip is the first candidate to sustain the extra plastic strain during plastic deformation, which is consistent with the conclusion made by Styczynski et al. [28] that the pyramidal <c+a> slip is a "compatibility first" deformation mechanism of the AZ31 magnesium alloy. As for the case of combination B, the activity of pyramidal <c+a> slip increases obviously during IPT along TD and further results in the splitting of pole densities around ND in the $\{0002\}$ pole figure. However, during IPC along RD, the increment with respect to the activity of pyramidal <c+a> slip is small, hence its influence on the predicted texture is very small. In the case of combination C, during IPC along RD, the appearance of pole densities close to ND can be attributed to the increase of activity in the $\{10\bar{1}1\}$ contraction twinning, which mainly results in the 56° rotation about c-axis in the crystals of AZ31 magnesium alloy. Moreover, the splitting of pole densities around RD can be due to the increasing activity in the prismatic <a> slip, as Chapuis et al. [24] have found that the prismatic <a> slip can generate a rotation around its c-axis in HCP metals. In terms of combination D, it can be concluded that whether it exists or not, the $\{10\bar{1}1\}$ contraction twinning would not affect the predicted textures during IPC along RD and IPT along TD.

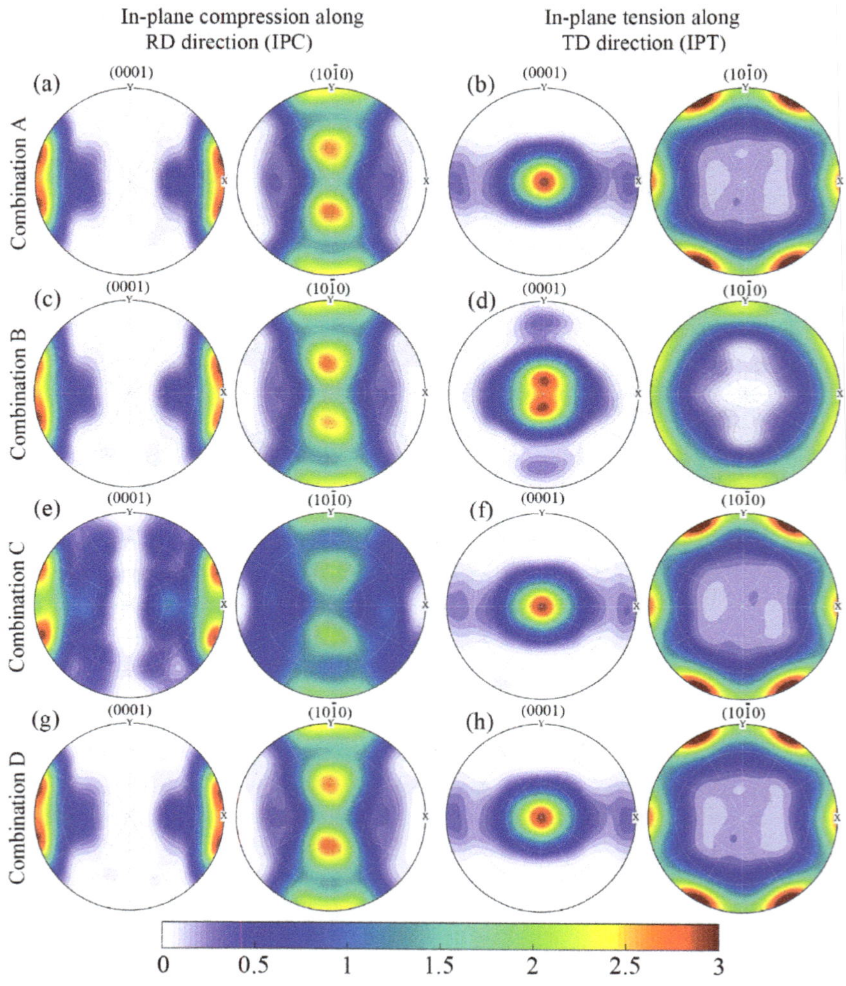

Figure 4. The predicted textures of the AZ31 magnesium alloy under uniaxial compression along RD and uniaxial tension along TD at the true strain of 0.15 based on various combinations of deformation mechanisms including (**a**,**b**) combination A; (**c**,**d**) combination B; (**e**,**f**) combination C; and (**g**,**h**) combination D.

4. Conclusions

The effect of secondary deformation mechanisms (prismatic <a> slip, pyramidal <c+a> slip, and $\{10\bar{1}1\}$ contraction twinning) on the plastic deformation of AZ31 magnesium alloy during in-plane compression (IPC) along the rolling direction (RD) and in-plane tension (IPT) along the transverse direction (TD) had been numerically investigated by VPSC modeling with different combinations of deformation mechanisms. The following conclusions can be drawn.

(1) As for the pyramidal <c+a> slip, it is the first candidate for sustaining the extra plastic strain during the plastic deformation. The increment in the activity of the pyramidal <c+a> slip contributes to the increase of flow stress in mechanical response during IPC along RD and IPT along TD. Moreover, during IPT along TD, its great growth contributes to the splitting of pole densities around ND in the $\{0002\}$ pole figure.

(2) The increment in the activity of prismatic <a> slip is also conducive to the improvement of flow stress in mechanical response during IPT along TD. In addition, during IPC along RD, its great growth contributes to the splitting of pole densities around RD in the $\{0002\}$ pole figure.

(3) The $\{10\overline{1}1\}$ contraction twinning has a minor influence on the plastic deformation of AZ31 magnesium alloy during IPC along RD and IPT along TD as well as the corresponding mechanical responses. However, during IPC along RD, its great growth contributes to the appearance of pole densities close to ND. Consequently, it is better to include the $\{10\overline{1}1\}$ contraction twinning in VPSC modeling to predict the texture more accurately during the plastic deformation of AZ31 magnesium alloy.

Author Contributions: Conceptualization, methodology, visualization, and writing—original draft preparation, Y.L.; software, validation, and writing—review and editing, L.H.; formal analysis and data curation, T.Z. and M.Y.; and supervision, J.Z.

Funding: This research was funded by the National Key Research and Development Program of China (grant No. 2016YFB0301105).

Conflicts of Interest: The authors declare no conflict of interest.

References

1. Wang, Z.Q.; Chapuis, A.; Qing, L. Simulation of mechanical behavior of AZ31 magnesium alloy during twin-dominated large plastic deformation. *Trans. Nonferr. Met. Soc. China* **2015**, *25*, 3595–3603. [CrossRef]
2. Lentz, M.; Risse, M.; Schaefer, N.; Reimers, W.; Beyerlein, I. Strength and ductility with {1011}—{1012} double twinning in a magnesium alloy. *Nat. Commun.* **2016**, *7*, 11068. [CrossRef] [PubMed]
3. Kabirian, F.; Khan, A.S.; Gnäupel-Herlod, T Visco-plastic modeling of mechanical responses and texture evolution in extruded AZ31 magnesium alloy for various loading conditions. *Int. J. Plast.* **2015**, *68*, 1–20. [CrossRef]
4. Pandey, A.; Kabirian, F.; Hwang, J.H.; Choi, S.H.; Khan, A.S. Mechanical responses and deformation mechanisms of an AZ31 Mg alloy sheet under dynamic and simple shear deformations. *Int. J. Plast.* **2015**, *68*, 111–131. [CrossRef]
5. Agnew, S.; Yoo, M.; Tome, C. Application of texture simulation to understanding mechanical behavior of mg and solid solution alloys containing li or y. *Acta Mater.* **2001**, *49*, 4277–4289. [CrossRef]
6. Agnew, S.R.; Duygulu, Ö. Plastic anisotropy and the role of non-basal slip in magnesium alloy AZ31B. *Int. J. Plast.* **2005**, *21*, 1161–1193. [CrossRef]
7. Koike, J.; Kobayashi, T.; Mukai, T.; Watanabe, H.; Suzuki, M.; Maruyama, K.; Higashi, K. The activity of non-basal slip systems and dynamic recovery at room temperature in fine-grained AZ31B magnesium alloys. *Acta Mater.* **2003**, *51*, 2055–2065. [CrossRef]
8. Lévesque, J.; Inal, K.; Neale, K.W.; Mishra, R. Numerical modeling of formability of extruded magnesium alloy tubes. *Int. J. Plast.* **2010**, *26*, 65–83. [CrossRef]
9. Kim, J.H.; Kim, D.; Lee, Y.S.; Lee, M.G.; Chung, K.; Kim, H.Y.; Wagoner, R.H. A temperature-dependent elasto-plastic constitutive model for magnesium alloy AZ31 sheets. *Int. J. Plast.* **2013**, *50*, 66–93. [CrossRef]
10. Wang, H.; Wu, P.; Wang, J.; Tomé, C. A crystal plasticity model for hexagonal close packed (hcp) crystals including twinning and de-twinning mechanisms. *Int. J. Plast.* **2013**, *49*, 36–52. [CrossRef]
11. Ardeljan, M.; Beyerlein, I.J.; Knezevic, M. Effect of dislocation density-twin interactions on twin growth in AZ31 as revealed by explicit crystal plasticity finite element modeling. *Int. J. Plast.* **2017**, *99*, 81–101. [CrossRef]
12. Clayton, J.D.; Knap, J. A phase field model of deformation twinning: Nonlinear theory and numerical simulations. *Phys. D Nonlinear Phenom.* **2011**, *240*, 841–858. [CrossRef]
13. Staroselsky, A.; Anand, L. A constitutive model for hcp materials deforming by slip and twinning: Application to magnesium alloy AZ31B. *Int. J. Plast.* **2003**, *19*, 1843–1864. [CrossRef]

14. Lebensohn, R.A.; Tomé, C. A self-consistent anisotropic approach for the simulation of plastic deformation and texture development of polycrystals: Application to zirconium alloys. *Acta Metall. Mater.* **1993**, *41*, 2611–2624. [CrossRef]
15. Eshelby, J.D. The determination of the elastic field of an ellipsoidal inclusion, and related problems. *Proc. R. Soc. Lond. A* **1957**, *241*, 376–396.
16. Wang, H.; Raeisinia, B.; Wu, P.; Agnew, S.; Tomé, C. Evaluation of self-consistent polycrystal plasticity models for magnesium alloy AZ31B sheet. *Int. J. Solids Struct.* **2010**, *47*, 2905–2917. [CrossRef]
17. Lebensohn, R.; Gonzalez, M.; Tomé, C.; Pochettino, A. Measurement and prediction of texture development during a rolling sequence of zircaloy-4 tubes. *J. Nucl. Mater.* **1996**, *229*, 57–64. [CrossRef]
18. Li, H.; Zhang, H.; Yang, H.; Fu, M.; Yang, H. Anisotropic and asymmetrical yielding and its evolution in plastic deformation: Titanium tubular materials. *Int. J. Plast.* **2017**, *90*, 177–211. [CrossRef]
19. Mandal, S.; Gockel, B.T.; Balachandran, S.; Banerjee, D.; Rollett, A.D. Simulation of plastic deformation in Ti-5553 alloy using a self-consistent viscoplastic model. *Int. J. Plast.* **2017**, *94*, 57–73. [CrossRef]
20. Rauch, E.; Gracio, J.; Barlat, F.; Vincze, G. Modelling the plastic behaviour of metals under complex loading conditions. *Model. Simul. Mater. Sci. Eng.* **2011**, *19*, 035009. [CrossRef]
21. Segurado, J.; Lebensohn, R.A.; LLorca, J.; Tomé, C.N. Multiscale modeling of plasticity based on embedding the viscoplastic self-consistent formulation in implicit finite elements. *Int. J. Plast.* **2012**, *28*, 124–140. [CrossRef]
22. Zhou, G.; Jain, M.K.; Wu, P.; Shao, Y.; Li, D.; Peng, Y. Experiment and crystal plasticity analysis on plastic deformation of AZ31B Mg alloy sheet under intermediate temperatures: How deformation mechanisms evolve. *Int. J. Plast.* **2016**, *79*, 19–47. [CrossRef]
23. Tomé, C.; Lebensohn, R.A.; Kocks, U. A model for texture development dominated by deformation twinning: Application to zirconium alloys. *Acta Metall. Mater.* **1991**, *39*, 2667–2680. [CrossRef]
24. Chapuis, A.; Wang, Z.Q.; Liu, Q. Influence of material parameters on modeling plastic deformation of mg alloys. *Mater. Sci. Eng. A* **2016**, *655*, 244–250. [CrossRef]
25. Hu, L.; Jiang, S.; Zhou, T.; Chen, Q. A coupled finite element and crystal plasticity study of friction effect on texture evolution in uniaxial compression of niti shape memory alloy. *Materials* **2018**, *11*, 2162. [CrossRef] [PubMed]
26. Lebensohn, R.; Tomé, C. *Manual for Code Visco-Plastic Self-Consistent (Vpsc), Version 7b*; Los Alamos National Laboratory: Los Alamos, NM, USA, 2007.
27. Oppedal, A.; El Kadiri, H.; Tomé, C.; Kaschner, G.; Vogel, S.C.; Baird, J.; Horstemeyer, M. Effect of dislocation transmutation on modeling hardening mechanisms by twinning in magnesium. *Int. J. Plast.* **2012**, *30*, 41–61. [CrossRef]
28. Styczynski, A.; Hartig, C.; Bohlen, J.; Letzig, D. Cold rolling textures in AZ31 wrought magnesium alloy. *Scr. Mater.* **2004**, *50*, 943–947. [CrossRef]

metals

MDPI

Article

Investigation of the Dynamic Recrystallization of FeMnSiCrNi Shape Memory Alloy under Hot Compression Based on Cellular Automaton

Yu Wang [1,2], Xiaodong Xing [1,*], Yanqiu Zhang [1] and Shuyong Jiang [1,*]

1 College of Mechanical and Electrical Engineering, Harbin Engineering University, Harbin 150001, China; wangyuhrbeu@126.com (Y.W.); zhangyq@hrbeu.edu.cn (Y.Z.)

2 College of Materials Science and Chemical Engineering, Harbin Engineering University, Harbin 150001, China

* Correspondence: xingxiaodong@hrbeu.edu.cn (X.X.); jiangshuyong@hrbeu.edu.cn (S.J.); Tel.: +86-451-82519710 (X.X. & S.J.)

Received: 29 March 2019; Accepted: 19 April 2019; Published: 22 April 2019

Abstract: Dynamic recrystallization (DRX) takes place when FeMnSiCrNi shape memory alloy (SMA) is subjected to compression deformation at high temperatures. Cellular automaton (CA) simulation was used for revealing the DRX mechanism of FeMnSiCrNi SMA by predicting microstructures, grain size, flow stress, and dislocation density. The DRX of FeMnSiCrNi SMA has a characteristic of repeated nucleation and finite growth. The size of recrystallized grains increases with increasing deformation temperatures, but it decreases with increasing strain rates. The increase of deformation temperature leads to the decrease of the flow stress, whereas the increase in strain rate results in the increase of the flow stress. The dislocation density exhibits the same situation as the flow stress. The simulated results were supported by the experimental ones very well. Dislocation density is a crucial factor during DRX of FeMnSiCrNi SMA. It affects not only the nucleation but also the growth of the recrystallized grains. Occurrence of DRX depends on a critical dislocation density. The difference between the dislocation densities of the recrystallized and original grains becomes the driving force for the growth of the recrystallized grains, which lays a solid foundation for the recrystallized grains growing repeatedly.

Keywords: FeMnSiCrNi alloy; shape memory alloy; cellular automaton; dynamic recrystallization

1. Introduction

Much attention has been paid to FeMnSi shape memory alloys (SMAs) since they were discovered because they possess low manufacturing costs, good formability, high mechanical properties, etc. [1,2]. The shape memory effect of FeMnSi SMAs stems from the transformation of γ austenite to ε martensite induced by external stress, where the corresponding crystal structure is changed from face-centered cubic (FCC) to close-packed hexagonal (HCP) structure [3–5]. Many researchers have devoted themselves to adding the alloying elements based on the FeMnSi SMAs in order to further enhance the shape memory effect or mechanical properties. The alloying elements deal with Cr, Ni, Sm, Ta, Nb, and Ti [6–9]. Among the alloying elements, the addition of Cr and Ni contributes to improving the corrosion resistance of FeMnSi SMAs. The FeMnSiCrNi SMAs have been widely investigated by many researchers [10–12]. It is generally accepted that hot working is an indispensable approach to making FeMnSiCrNi SMAs into products for engineering application. In particular, it is of great importance to reveal the DRX mechanism of FeMnSiCrNi SMAs because DRX microstructures have an important influence on the transformation behavior of FeMnSiCrNi SMAs.

It is well known that cellular automaton (CA) is a typical algorithm that is able to represent the discrete spatial and temporal evolution of complex system, where a local or global transformation rule is used for the involved cells [13,14]. In addition, the transformation rule, which is likely to be deterministic or probabilistic, is used to determine the state variables of a lattice point according to its previous state variables along with the state variables of the neighboring sites [15,16]. Therefore, CA has been an effective simulation tool to predict the microstructures of metal materials. In particular, many researchers have been dedicated to developing all kinds of CAs to be used for describing DRX behavior of metals, including steels [17,18], magnesium alloy [19–21], Ni-based superalloy [22], and vanadium alloy [23]. The CA model is able to succeed in clarifying the nucleation and growth of the recrystallized grains during DRX of metal materials.

In the current study, a CA model is developed to understand the DRX behavior of FeMnSiCrNi SMAs undergoing plastic deformation at elevated temperature. The simulation results are supported by the experimental ones as well. The involved contents have not been reported in the literature until now.

2. Materials and Methods

The as-received Fe66Mn15Si5Cr9Ni5 (wt.%) SMA was used in the present study and it is simply termed as FeMnSiCrNi alloy. The FeMnSiCrNi samples, which possess a diameter of 6 mm and a height of 9 mm, were fabricated from the as-received FeMnSiCrNi alloy in order to be used for the compression experiments on the AG-Xplus equipment (Shimadzu Corporation, Hadano, Japan). In the compression experiments, the strain rates selected were 0.005, 0.05, and 0.5 s^{-1}, and the temperatures determined were 850, 900, 950, and 1000 °C, respectively. In addition, the compressed deformation degree was 60%, which corresponds to the true strain of 0.9. The FeMnSiCrNi samples were firstly put into the heating furnace of the AG-Xplus equipment and then were heated to the desired temperatures, followed by being held for two minutes. Subsequently, they were subjected to compression and then were immediately put into water at room temperature. The compressed FeMnSiCrNi samples were cut along the compressive direction to observe the microstructures of the axial section. The FeMnSiCrNi specimens for optical microstructure analysis were firstly subjected to coarse grinding by means of abrasive paper with 1500 meshes and then underwent a fine grind using abrasive paper with 3000 meshes. Subsequently, they were polished by polishing cloth. Finally, they were etched in a solution with 6.7% $CuSO_4$ + 26.3% HCl + 67% H_2O, and the corresponding microstructures were characterized using a LEICA DM IRM optical microscope (Leica Camera AG, Solms, Germany).

3. Fundamentals of CA modeling

3.1. Model of Dislocation Evolution

The two competitive processes, namely, work hardening and DRX, coexist during plastic deformation of FeMnSiCrNi SMA at high temperatures. Consequently, the two processes lead to the variation of dislocation density. For clarifying the deformation characteristic of FeMnSiCrNi SMA, the Kocks-Mecking (K-M) model was adopted in the present work, where dislocation density is associated with the strain according to the following equation [24].

$$\frac{d\rho}{d\varepsilon} = k_1 \sqrt{\rho} - k_2 \rho \quad (1)$$

where ϵ is true strain, k_1 represents the parameter responsible for work hardening and k_2 stands for the parameter responsible for softening. In addition, k_1 and k_2 are expressed by the following two Equations [25].

$$k_1 = 2\theta_0 / \alpha' G b \quad (2)$$

$$k_2 = 2\theta_0 / \sigma_s \quad (3)$$

where σ_s is saturation stress, α' represents interaction between dislocations, whose value generally is 0.5–1.0, G is the shear modulus, depending on Young's modulus E as well as Poisson's ratio μ, b is Burgers vector, and θ_0 stands for work-hardening rate and it is solved by the following Equation [25].

$$\frac{d\sigma}{d\varepsilon} = \theta_0\left(1 - \frac{\sigma}{\sigma_s}\right) \tag{4}$$

where σ is flow stress, which is associated with dislocation density. Consequently, it satisfies the following equation [24,26].

$$\sigma = \alpha' Gb\sqrt{\overline{\rho}} \tag{5}$$

where $\overline{\rho}$ is the average dislocation density, and it can be expressed by the dislocation density of each cell ρ_i, namely,

$$\overline{\rho} = \frac{1}{N}\sum_{i=1}^{n} \rho_i \tag{6}$$

where N represents the total cell number.

3.2. Model for Nucleation Rate

The successive nucleation model used in the present study was established by Ding and Guo [27]. According to their model, nucleation rate $\dot{n}$ depends on deformation temperature T as well as strain rate $\dot{\varepsilon}$, namely,

$$\dot{n}(\dot{\varepsilon}, T) = C\dot{\varepsilon}^{m}\exp(-\frac{Q_{act}}{RT}) \tag{7}$$

where C is a constant, R represents gas constant, m stands for material constant and it equals 1.0 based on References [25,28], and Q_{act} is the activation energy. Therefore, the constitutive behavior of FeMnSiCrNi SMA during plastic deformation at high temperatures satisfies the Arrhenius equation [13], namely,

$$\dot{\varepsilon} = A_0[\sinh(\alpha\sigma)]^{n}\exp(-\frac{Q_{act}}{RT}) \tag{8}$$

where A_0, α, and n are the material constants. According to the mathematical transformation of Equation (8), Q_{act} is able to be expressed as follows [29].

$$Q_{act} = R\left[\frac{\partial \ln\dot{\varepsilon}}{\partial \ln\sinh(\alpha\sigma)}\right]_T\left[\frac{\partial \ln\sinh(\alpha\sigma)}{\partial(1/T)}\right]_{\dot{\varepsilon}} \tag{9}$$

The constant C in Equation (7) is able to be solved based on the experimental data. After the DRX percentage η is measured by experiment, $\dot{n}$ is calculated as follows [25].

$$\eta = \dot{n}\frac{\varepsilon}{\dot{\varepsilon}}\frac{4}{3}\pi r_d^3 \tag{10}$$

where r_d represents average radius of recrystallized grains, and it is calculated as follows [30].

$$\frac{\sigma}{G}\left(\frac{2r_d}{b}\right)^{n'} = K \tag{11}$$

where n' is approximately 2/3, and K is about 10 for most metals.

The critical dislocation density, where DRX begins to occur, is expressed by the following Equation [31].

$$\rho_c = \left(\frac{20\gamma\dot{\varepsilon}}{3blM\tau^2}\right)^{1/3} \tag{12}$$

where τ is dislocation line energy, M represents grain boundary mobility ratio, l stands for mean free path of dislocation, and γ is the grain boundary energy. Furthermore, they are capable of be calculated by Equations (13)–(16), respectively [24].

$$\tau = c_2 G b^2 \tag{13}$$

where c_2 is a constant, and it is determined as 0.5.

$$M = \frac{\delta D_{\mathrm{ob}} b}{kT} \exp\left(\frac{-Q_{\mathrm{b}}}{RT}\right) \tag{14}$$

where δ represents characteristic thickness of grain boundary, D_{ob} represents boundary self-diffusion coefficient, k stands for Boltzmann' s constant, and Q_b represents activation energy of boundary diffusion.

$$l = \frac{KGb}{\sigma} \tag{15}$$

$$\gamma_i = \begin{cases} \gamma_{\mathrm{m}} & \theta_i \geq 15^\circ \\ \gamma_i = \gamma_{\mathrm{m}} \frac{\theta_i}{\theta_{\mathrm{m}}}\left(1 - \ln\left(\frac{\theta_i}{\theta_{\mathrm{m}}}\right)\right) & \theta_i < 15^\circ \end{cases} \tag{16}$$

where θ_i represents misorientation between the ith recrystallized grain and its adjacent one, θ_{m} and γ_{m} represent grain boundary misorientation and grain boundary energy, respectively, when a high angle boundary (greater than 15°) occurs. In addition, γ_{m} is calculated as follows [32].

$$\gamma_{\mathrm{m}} = \frac{Gb\theta_{\mathrm{m}}}{4\pi(1 - \mu)} \tag{17}$$

3.3. Model for CA

In the present study, a two-dimensional model for CA was established in order to simulate the microstructural evolution during DRX of FeMnSiCrNi SMA. A 235 × 235 square lattice was used for implementing the involved simulation. One lattice corresponded to the size of 1 μm in the real material. Consequently, the simulation area was equivalent to 0.235 × 0.235 mm^2 in a genuine FeMnSiCrNi SMA sample. In addition, the periodic boundary condition was used to simulate an infinite space. The V. Neumann's neighboring rule was adopted. The involved simulation parameters are shown in Table 1. The other details with respect to the CA model can be found in literature [33].

Table 1. The parameters for Cellular automaton (CA) simulation in FeMnSiCrNi shape memory alloy (SMA).

Parameter	b (m)	G (MN·m^{-2})	Q_{act} (KJ·mol^{-1})	Q_b (KJ·mol^{-1})	δD_0 (m^3·s^{-1})
Value	2.58×10^{-10}	2.67×10^4	350	159	7.5×10^{-14}

4. Results and Discussion

4.1. Flow Behavior of FeMnSiCrNi SMA

Figure 1 demonstrates true stress–strain curves of FeMnSiCrNi SMA subjected to compressive deformation. It is obvious that the stress value of FeMnSiCrNi SMA increases with strain rate, whereas they decrease with deformation temperature. In particular, the true stress increases with increasing true strain at the beginning, which indicates that the increasing dislocation density leads to the occurrence of working hardening. When plastic deformation continues to take place, flow stress possesses a stable value with increasing true strain. This reveals that DRX takes place during plastic deformation of FeMnSiCrNi SMA.

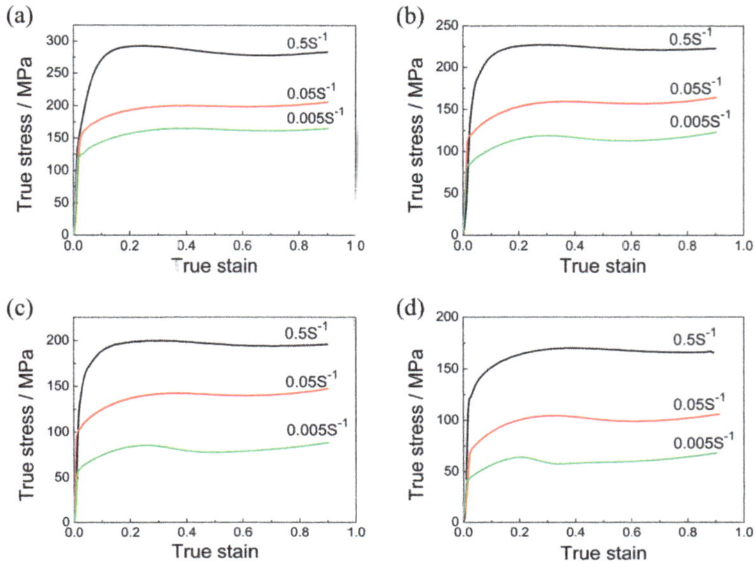

Figure 1. Compressive stress-strain curves of FeMnSiCrNi SMA undergoing compression: (**a**) $T = 850$ °C; (**b**) $T = 900$ °C; (**c**) $T = 950$ °C; (**d**) $T = 1000$ °C.

4.2. Microstructural Evolution of DRX

In order to reveal the mechanism of DRX, CA simulation was used to capture microstructural evolution of FeMnSiCrNi SMA subjected to compression, where the temperature and the strain rate were selected as 1000 °C and 0.005 s^{-1}, respectively. The corresponding simulation results are illustrated in Figure 2. It was noted that no new grains occur at the point B of stress-strain curve in Figure 2a, as illustrated in Figure 2b. The stage can be viewed as the elastic deformation zone, and thus no new dislocations were induced. With increasing plastic strain, plenty of dislocations are formed. When the dislocation density achieved a critical value, new grains nucleated at the grain boundary, which corresponded to point C of the stress-strain curve in Figure 2a, as shown Figure 2c. It is well known that DRX is able to repeatedly nucleate and finitely grow. For example, it can be observed from Figure 2d that the new grains arose in the dynamically recrystallized grains. With the repeated progression of DRX, almost complete DRX occurred, as shown Figure 2f.

4.3. Prediction of Microstructures

Figure 3 indicates microstructures of FeMnSiCrNi SMA undergoing compression at the various temperatures by virtue of CA simulation, where a strain rate of 0.5 s^{-1} as well as a true strain of 0.9 is chosen. It can be observed from Figure 3 that the size of dynamically recrystallized grains increases with temperature. In addition, the fraction of recrystallization increases with temperature as well. To validate the CA simulation results, the corresponding optical microstructures of FeMnSiCrNi SMA are characterized, as shown in Figure 4. It is very obvious that the size of grains also increases with temperature. The simulation results are supported by the experimental ones very well. In addition, it can be found from Figure 4 that martensite and twin occur in FeMnSiCrNi SMA. As Straumal et al. [34,35] stated, the twins frequently appear in FCC metals with low-to-medium stacking fault energy. Martensite and the twins are readily observed in FeMnSiCrNi SMA since FeMnSiCrNi SMA possesses low stacking fault energy [9–11]. The formation mechanisms of martensite and the twins in FeMnSiCrNi SMA are out of the scope of the manuscript and shall be investigated by means of further experimental evidence in the future.

Figure 5 shows microstructures of FeMnSiCrNi SMA undergoing compression at various strain rates at 850 °C via the CA model, which corresponds to a true strain of 0.9. In Figure 5, the size of recrystallized grains increases with strain rate, but the effect of strain rate is not as considerable as the effect of temperature. In the same manner, the simulation results are also supported by the experimental ones, as illustrated in Figure 6. It is very obvious that the slower strain rate provides more nucleation and growth time for DRX of FeMnSiCrNi SMA. However, the strain rate is only able to exert a slight impact on the growth of grains during DRX of FeMnSiCrNi SMA due to the limitation of temperature.

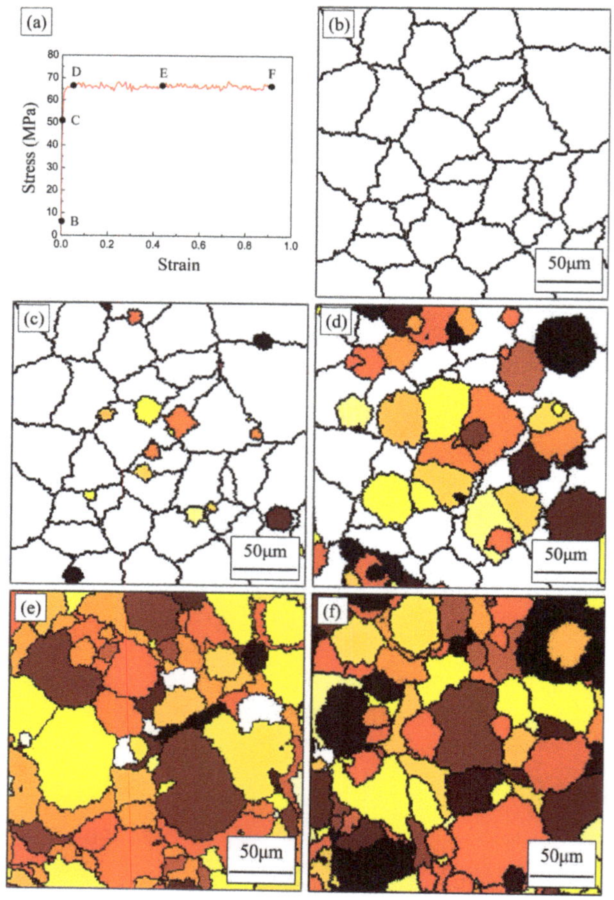

Figure 2. Microstructural evolution of FeMnSiCrNi SMA undergoing compression (T = 1000 °C, $\dot{\varepsilon}$ = 0.005 s^{-1}) based on CA simulation: (**a**) Stress-strain curve; (**b**) point B in (**a**); (**c**) point C in (**a**); (**d**) point D in (**a**); (**e**) point E in (**a**); (**f**) point F in (**a**).

Figure 7 shows microstructures of FeMnSiCrNi SMA undergoing compression at the various strain rates at 1000 °C in the CA model, which corresponds to a true strain of 0.9. It is noted in Figure 7 that the increase in strain rate leads to the increase in the size of recrystallized grains as well. However, with increasing deformation temperature, strain rate more significantly influences the size of recrystallized grains. Furthermore, the simulation results are also supported by the experimental ones, as shown in Figure 8. Therefore, it is of greater significance to comprehensively consider the influence of deformation temperature and strain rate on DRX of FeMnSiCrNi SMA.

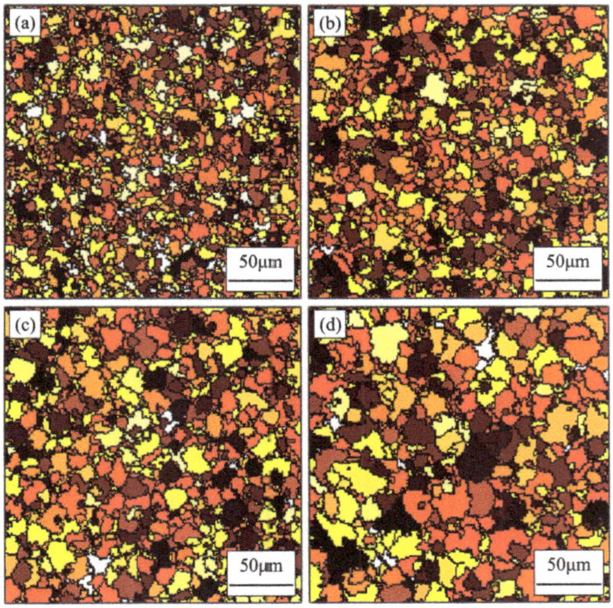

Figure 3. Microstructures of FeMnSiCrNi SMA undergoing compression ($\dot{\varepsilon} = 0.5\ s^{-1}$, $\varepsilon = 0.9$) based on CA simulation: (**a**) $T = 850$ °C; (**b**) $T = 900$ °C; (**c**) $T = 950$ °C; (**d**) $T = 1000$ °C.

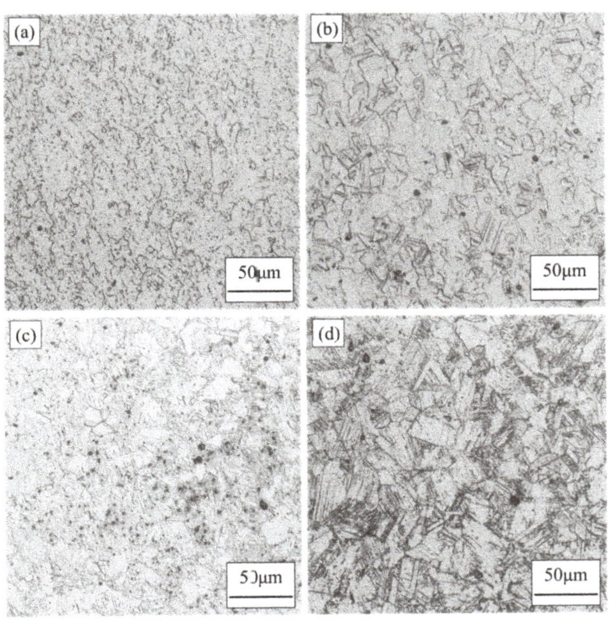

Figure 4. Optical microstructures of FeMnSiCrNi SMA undergoing compression ($\dot{\varepsilon} = 0.5\ s^{-1}$, $\varepsilon = 0.9$): (**a**) $T = 850$ °C; (**b**) $T = 900$ °C; (**c**) $T = 950$ °C; (**d**) $T = 1000$ °C.

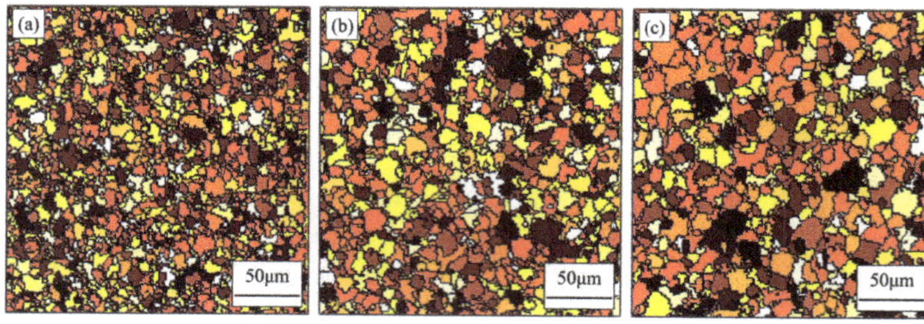

Figure 5. Microstructures of FeMnSiCrNi SMA undergoing compression (T = 850 °C, ε = 0.9) using CA simulation: (**a**) $\dot{\varepsilon} = 0.5\ \text{s}^{-1}$; (**b**) $\dot{\varepsilon} = 0.05\ \text{s}^{-1}$; (**c**) $\dot{\varepsilon} = 0.005\ \text{s}^{-1}$.

Figure 6. Optical Microstructures of FeMnSiCrNi SMA undergoing compression (T = 850 °C, ε = 0.9): (**a**) $\dot{\varepsilon} = 0.5\ \text{s}^{-1}$; (**b**) $\dot{\varepsilon} = 0.05\ \text{s}^{-1}$; (**c**) $\dot{\varepsilon} = 0.005\ \text{s}^{-1}$.

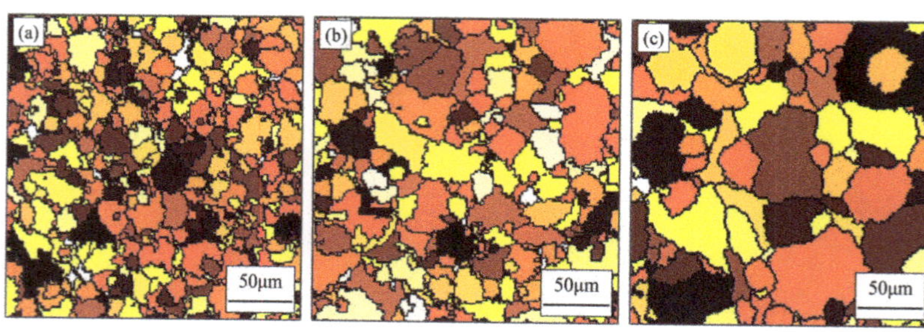

Figure 7. Microstructures of FeMnSiCrNi SMA undergoing compression (T = 1000 °C, ε = 0.9) based on CA simulation: (**a**) $\dot{\varepsilon} = 0.5\ \text{s}^{-1}$; (**b**) $\dot{\varepsilon} = 0.05\ \text{s}^{-1}$; (**c**) $\dot{\varepsilon} = 0.005\ \text{s}^{-1}$.

Figure 8. Optical microstructures of FeMnSiCrNi SMA undergoing compression (T = 1000 °C, ε = 0.9): (**a**) $\dot{\varepsilon} = 0.5\ \text{s}^{-1}$; (**b**) $\dot{\varepsilon} = 0.05\ \text{s}^{-1}$; (**c**) $\dot{\varepsilon} = 0.005\ \text{s}^{-1}$.

4.4. Prediction of Flow Stress

Figure 9 indicates the compressive stress–strain curves of FeMnSiCrNi SMA at the various temperatures and strain rates based on CA simulation. Comparing Figure 9 to Figure 1, obviously, the simulated results are supported by the experimental ones. However, it is not denied that there exists a certain difference between the simulating and experimental results except at 0.5 s^{-1}. In the case of the other two strain rates, in particular, a certain fluctuation appears in the stress-strain curves of FeMnSiCrNi SMA. The fluctuation is closely associated with evolution of dislocation density. This stems from the sharp competition between dynamic softening and work hardening. This also demonstrates that the present CA simulation possesses a certain limitation in predicting the flow behavior of FeMnSiCrNi SMA during compression deformation.

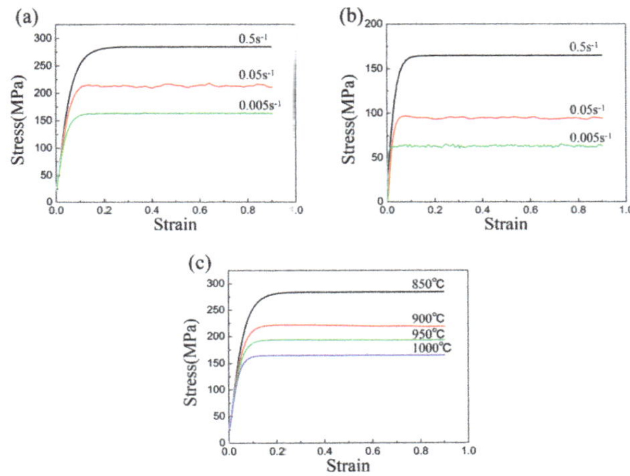

Figure 9. Compression stress-strain curves of FeMnSiCrNi SMA at the various temperatures and strain rates based on CA simulation: (**a**) $T = 850$ °C; (**b**) $T = 1000$ °C; (**c**) $\dot{\varepsilon} = 0.5$ s^{-1}.

4.5. Prediction of Dislocation Density

Generally, dislocation density a substantial factor during DRX of FeMnSiCrNi SMA. In other words, dislocation density is a crucial factor in terms of determining whether DRX of FeMnSiCrNi SMA takes place. The new crystal nucleus was formed only when a critical dislocation density was reached. Therefore, it was quite necessary to simulate the evolution of dislocation density via CA, which provides a theoretical foundation for clarifying the DRX mechanism of FeMnSiCrNi SMA. For the purpose, dislocation density of FeMnSiCrNi SMA corresponding to various deformation temperatures and strain rates was predicted by CA simulation, where a true strain of 0.9 was selected, as shown in Figure 10, Figure 11, and Figure 12. Figure 10 indicates a dislocation density of FeMnSiCrNi SMA corresponding to various deformation temperatures, where the strain rate was selected as 0.5 s^{-1}. It is very clear that the dislocation density was reduced with increasing deformation temperature. Furthermore, according to CA simulation, the dislocation density of FeMnSiCrNi SMA corresponding to various strain rates at 850 and 1000 °C is illustrated in Figures 11 and 12, respectively. It is evident that dislocation density increases with stain rate. Consequently, it is of great importance for interpreting DRX mechanism of FeMnSiCrNi SMA to comprehensively take into account effect of deformation temperature and strain rate on dislocation density. Dislocation density particularly exhibits a similar trend with flow stress. As is described in the aforementioned section, the occurrence of DRX is based on the fact that dislocation density is lower in the recrystallized grains than in the original ones. In other words, the difference of the two dislocation densities shall become the driving force to cause the recrystallized grains to grow, which lays the theoretical foundation for the recrystallized grains growing repeatedly.

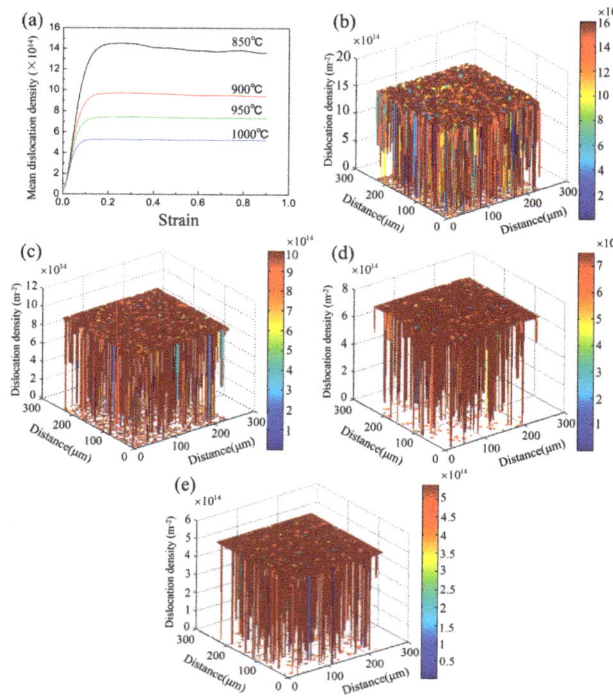

Figure 10. Dislocation density of FeMnSiCrNi SMA under dynamic recrystallization (DRX) at the various deformation temperatures at the strain rate of 0.5 s^{-1} based on CA: (**a**) variation of dislocation density with strain at the various temperatures; (**b**) $\varepsilon = 0.9$, 850 °C; (**c**) $\varepsilon = 0.9$, 900 °C; (**d**) $\varepsilon = 0.9$, 950 °C; (**e**) $\varepsilon = 0.9$, 1000 °C.

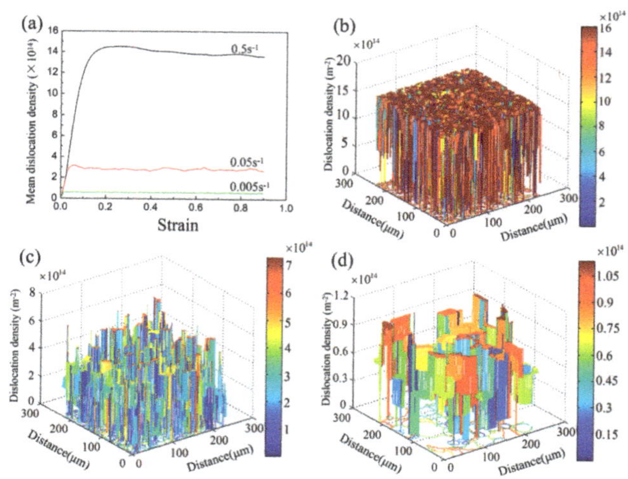

Figure 11. Dislocation density of FeMnSiCrNi SMA under DRX at the various strain rates at 850 °C based on CA: (**a**) variation of dislocation density with strain; (**b**) $\varepsilon = 0.9$, 0.5 s^{-1}; (**c**) $\varepsilon = 0.9$, 0.05 s^{-1}; (**d**) $\varepsilon = 0.9$, 0.005 s^{-1}.

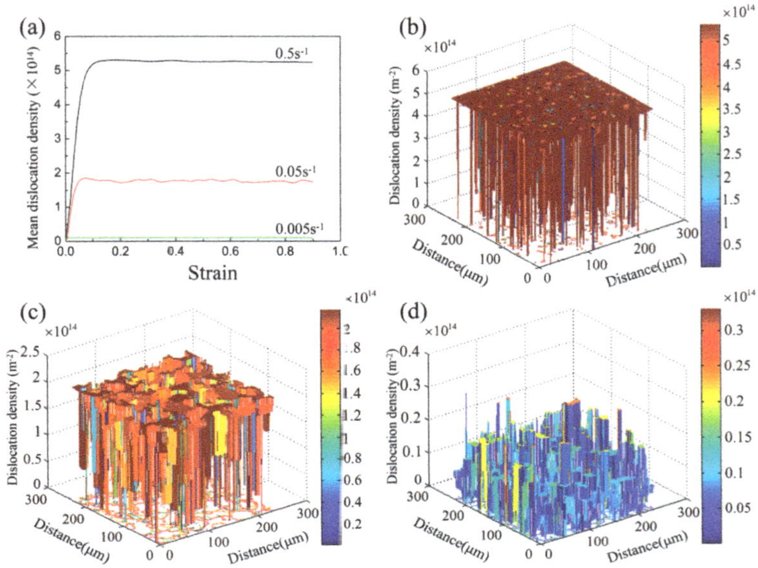

Figure 12. Dislocation density of FeMnSiCrNi SMA under DRX at the various strain rates at 1000 °C based on CA: (a) variation of dislocation density with strain; (b) $\varepsilon = 0.9$, 0.5 s^{-1}; (c) $\varepsilon = 0.9$, 0.05 s^{-1}; (d) $\varepsilon = 0.9$, 0.005 s^{-1}.

5. Conclusions

(1) Dynamic recrystallization (DRX) is found to appear during plastic deformation of FeMnSiCrNi SMA at elevated temperatures. Cecullar automaton (CA) simulation becomes an effect candidate for revealing the DRX mechanism of FeMnSiCrNi SMA by predicting microstructures, grain size, flow stress, and dislocation density. Dynamic recrystallization (DRX) of FeMnSiCrNi SMA has a characteristic of repeated nucleation and finite growth. The size of dynamically recrystallized grains increases with deformation temperature, but it decreases with strain rate. The increasing deformation temperature leads to the decrease of flow stress, whereas the decreasing strain rate results in the increase of flow stress. The dislocation density exhibits the same situation as the flow stress. The simulated results are supported by the experimental ones very well.

(2) Dislocation density plays a crucial role during DRX of FeMnSiCrNi SMA experiencing plastic deformation at high temperatures. It has an important influence on the nucleation as well as the growth of the dynamically recrystallized grains. The new crystal nucleus is generated only when a critical dislocation density is achieved. The difference between dislocation densities of the recrystallized and original grains shall become the driving force for causing the recrystallized grains to grow, which lays the solid foundation for the recrystallized grains growing repeatedly.

Author Contributions: Y.W. performed the experimental research and wrote the manuscript; X.X. compiled the cellular automaton program; Y.Z. developed cellular automaton software; S.J. supervised the manuscript.

Funding: The work was financially supported by the National Natural Science Foundation of China (Nos. 51475101 and 51871070).

Conflicts of Interest: The authors declare no conflict of interest.

References

1. Yang, C.F.; Lin, H.C.; Lin, K.M.; Tsai, H.K. Effect of thermo-mechanical treatment on a Fe-30Mn-6Si shape memory alloy. *Mater. Sci. Eng. A* **2008**, *497*, 445–450. [CrossRef]
2. Amini, R.; Shamsipoor, A.; Ghaffari, M.; Alizadeh, M.; Okyay, O.K. Phase transformation during mechano-synthesis of nanocrystalline/amorphous Fe-32Mn-6Si alloys. *Mater. Charact.* **2013**, *84*, 169–174. [CrossRef]
3. Stanford, N.; Dunne, D.P. Effect of Si on the reversibility of stress-induced martensite in Fe-Mn-Si shape memory alloy. *Acta Mater.* **2010**, *58*, 6752–6762. [CrossRef]
4. Caenegem, N.N.; Duprez, L.; Verbeken, K.; Segers, D.; Houbaert, Y. Stresses related to the shape memory effect in Fe-Mn-Si-based shape memory alloys. *Mater. Sci. Eng. A* **2008**, *481*, 183–189. [CrossRef]
5. Stanford, N.; Dunne, D.P. Thermo-mechanical processing and the shape memory effect in an Fe-Mn-Si-based shape memory alloy. *Mater. Sci. Eng. A* **2006**, *422*, 352–359. [CrossRef]
6. Shakoor, R.A.; Khalid, F.A. Thermomechanical behavior of Fe-Mn-Si-Cr-Ni shape memory alloys modified with samarium. *Mater. Sci. Eng. A* **2009**, *499*, 411–414. [CrossRef]
7. Lin, H.C.; Wu, S.K.; Peng, Y.T.; Cheng, T.C.; Lin, K.M. Pseudoelasticity of thermo-mechanically treated Fe-Mn-Si-Cr-Ta alloys. *J. Alloy Compd.* **2013**, *577*, S338–S343.
8. Stanford, N.; Dunne, D.P.; Li, H. Re-examination of effect of NbC precipitation on shape memory in Fe-Mn-Si-based alloys. *Scr. Mater.* **2008**, *58*, 583–586. [CrossRef]
9. Wang, G.X.; Peng, H.B.; Sun, P.P.; Wang, S.L.; Wen, Y.H. Effect of titanium addition on shape memory effect and recovery stress of training-free cast Fe-Mn-Si-Cr-Ni shape memory alloys. *Mater. Sci. Eng. A* **2016**, *657*, 339–346. [CrossRef]
10. Zhang, W.; Jiang, L.Z.; Li, N.; Wen, Y.H. Improvement of shape memory effect in an Fe-Mn-Si-Cr-Ni alloy fabricated by equal channel angular pressing. *J. Mater. Process. Technol.* **2008**, *208*, 130–134.
11. Druker, A.V.; Baruj, A.; Isola, L.; Fuster, V.; Malarría, J.; Bolmaro, R. Gaining flexibility in the design of microstructure, texture and shape memory properties of an Fe-Mn-Si-Cr-Ni alloy processed by ECAE and annealing. *Mater. Des.* **2016**, *107*, 153–162. [CrossRef]
12. Kwon, E.P.; Fujieda, S.; Shinoda, K.; Suzuki, S. Texture evolution and fcc/hcp transformation in Fe-Mn-Si-Cr-Ni alloys by tensile deformation. *Mater. Sci. Eng. A* **2010**, *527*, 6524–6532. [CrossRef]
13. Jin, Z.Y.; Cui, Z.S. Investigation on dynamic recrystallization using a modified cellular automaton. *Comput. Mater. Sci.* **2012**, *63*, 249–255. [CrossRef]
14. Timoshenkov, A.; Warczok, P.; Albu, M.; Klarner, J.; Kozeschnik, E.; Bureau, R.; Sommitsch, C. Modelling the dynamic recrystallization in C-Mn micro-alloyed steel during thermo-mechanical treatment using cellular automata. *Comput. Mater. Sci.* **2014**, *94*, 85–94. [CrossRef]
15. Popova, E.; Staraselski, Y.; Brahme, A.; Mishra, R.K.; Inal, K. Coupled crystal plasticity-Probabilistic cellular automata approach to model dynamic recrystallization in magnesium alloys. *Int. J. Plast.* **2015**, *66*, 85–102. [CrossRef]
16. Hallberg, H.; Wallin, M.; Ristinmaa, M. Simulation of discontinuous dynamic recrystallization in pure Cu using a probabilistic cellular automaton. *Comput. Mater. Sci.* **2010**, *49*, 25–34. [CrossRef]
17. Chen, M.S.; Yuan, W.Q.; Lin, Y.C.; Li, H.B.; Zou, Z.H. Modeling and simulation of dynamic recrystallization behavior for 42CrMo steel by an extended cellular automaton method. *Vacuum* **2017**, *146*, 142–151. [CrossRef]
18. Chen, F.; Qi, K.; Cui, Z.S.; Lai, X.M. Modeling the dynamic recrystallization in austenitic stainless steel using cellular automaton method. *Comput. Mater. Sci.* **2014**, *83*, 331–340. [CrossRef]
19. Chen, M.S.; Yuan, W.Q.; Li, H.B.; Zou, Z.H. Modeling and simulation of dynamic recrystallization behaviors of magnesium alloy AZ31B using cellular automaton method. *Comput. Mater. Sci.* **2017**, *136*, 163–172. [CrossRef]
20. Wang, L.X.; Fang, G.; Qian, L.G. Modeling of dynamic recrystallization of magnesium alloy using cellular automata considering initial topology of grains. *Mater. Sci. Eng. A* **2018**, *711*, 268–283. [CrossRef]
21. Li, X.L.; Li, X.L.; Zhou, H.T.; Li, F.B.; Liu, Q. Simulation of dynamic recrystallization in AZ80 magnesium alloy using cellular automaton. *Comput. Mater. Sci.* **2017**, *140*, 95–104.
22. Liu, Y.X.; Lin, Y.C.; Li, H.B.; Wen, D.X.; Chen, X.M.; Chen, M.S. Study of dynamic recrystallization in a Ni-based superalloy by experiments and cellular automaton model. *Mater. Sci. Eng. A* **2015**, *626*, 432–440. [CrossRef]

23. Cao, Z.H.; Sun, Y.; Wan, Z.P.; Yang, W.H.; Ren, L.L.; Hu, L.X. Cellular automaton simulation of dynamic recrystallization behavior in V-10Cr-5Ti alloy under hot deformation conditions. *Trans. Nonferrous Met. Soc. China* **2019**, *29*, 98–111. [CrossRef]
24. Ding, R.; Guo, Z.X. Microstructural modelling of dynamic recrystallisation using an extended cellular automaton approach. *Comput. Mater. Sci.* **2002**, *23*, 209–218. [CrossRef]
25. Kugler, G.; Turk, R. Modeling the dynamic recrystallization under multi-stage hot deformation. *Acta Mater.* **2004**, *52*, 4659–4668. [CrossRef]
26. Mecking, H.; Kocks, U.F. Kinetics of flow and strain-hardening. *Acta Metall.* **1981**, *29*, 1865–1875. [CrossRef]
27. Ding, R.; Guo, Z.X. Coupled quantitative simulation of microstructural evolution and plastic flow during dynamic recrystallization. *Acta Mater.* **2001**, *49*, 3163–3175. [CrossRef]
28. Mcqueen, H.J.; Ryan, N.D. Constitutive analysis in hot working. *Mater. Sci. Eng. A* **2002**, *322*, 43–63. [CrossRef]
29. Derby, B. Dynamic recrystallisation: The steady state grain size. *Scr. Metall. Mater.* **1992**, *27*, 1581–1585. [CrossRef]
30. Sandstorm, R.; Lagneborg, R. A model for hot working occurring by recrystallization. *Acta Metall.* **1975**, *23*, 387–398. [CrossRef]
31. Yazdipour, N.; Davies, C.H.J.; Hodgson, P.D. Microstructural modeling of dynamic recrystallization using irregular cellular automata. *Mater. Sci.* **2008**, *44*, 566–577. [CrossRef]
32. Chen, F.; Cui, Z.S.; Liu, J.; Chen, W.; Chen, S.J. Mesoscale simulation of the high-temperature austenitizing and dynamic recrystallization by coupling a cellular automaton with a topology deformation technique. *Mater. Sci. Eng. A* **2010**, *527*, 5539–5549. [CrossRef]
33. Zhang, Y.Q.; Jiang, S.Y.; Liang, Y.L.; Hu, L. Simulation of dynamic recrystallization of NiTi shape memory alloy during hot compression deformation based on cellular automaton. *Comput. Mater. Sci.* **2013**, *71*, 124–134. [CrossRef]
34. Straumal, B.B.; Polyakov, S.A.; Mittemeijer, E.J Temperature inflfluence on the faceting of $\sum3$ and $\sum9$ grain boundaries in Cu. *Acta Mater.* **2006**, *54*, 167–172. [CrossRef]
35. Straumal, B.B.; Kogtenkova, O.A.; Gornakova, A.S.; Sursaeva, V.G.; Baretzky, B. Review: Grain boundary faceting–roughening phenomena. *J. Mater. Sci.* **2016**, *51*, 382–404. [CrossRef]

metals

MDPI

Article

Springback Prediction of a Hot Stamping Component Based on the Area Fractions of Phases

Xiangji Li [1,2], Xu Yan [1] and Zhiqiang Zhang [1,*]

[1] Key Laboratory of Automobile Materials, Ministry of Education and College of Materials Science and Engineering, Jilin University, Changchun 130025, China; xjli@jlu.edu.cn (X.L.); yanxu_jlu@sina.com (X.Y.)

[2] Rolling Forging Research Institute, Jilin University, Changchun 130025, China

* Correspondence: zhangzq@jlu.edu.cn; Tel./Fax: +86-431-85334969

Received: 31 May 2019; Accepted: 14 June 2019; Published: 20 June 2019

Abstract: Different from traditional hot stamping components with full martensite, the new tailored hot stamping (THS) components have different quenched microstructures, which results in their lower shape accuracy. To investigate the influence of different quenched phases on the springback of a component, a THS experiment of a U-shaped component was performed with segmented heating and a cooling tool. The area fractions of phases at different tool temperatures were obtained by a two-stage color tint etching procedure. Results showed that the quenched phase of the cold zone was almost full martensite. The area fraction of martensite in the hot zone was reduced to the lowest 13% at the tool temperature of 600 °C, while the bainite content reached the highest at 70%. The springback angles at different tool temperatures for quenching were measured by 3D scanning technology and the reverse modeling method. It was revealed that the springback angle increased with the increase of martensite and yet decreased with the increase of bainite. The relationship between the springback angle and the area fractions of the quenched phases was established by means of multiple linear regression analyses. The error analysis results of the predictions and measurements showed that the springback analysis model, based on the area fractions of quenched phases, could be used to predict the springback of hot stamping components with tailored properties.

Keywords: boron steel; tailored hot stamping; phase transition; springback

1. Introduction

The demand for vehicle weight reduction and improved crash performance has led to the application of hot stamping of boron steel [1,2]. However, the components with high strength and low levels of ductility did not show a good performance in crash energy absorption [3,4]. To improve the energy absorption, tailored hot stamping (THS) components with tailored mechanical properties have been developed [5–8]. From continuous cooling transformation (CCT) curve, it can be seen that ferrite and bainite can be obtained at cooling rates lower than 30 °C/s [9,10]. Since the temperatures of two contact bodies will affect their cooling rates, THS with segmented heating and cooling tools can be used to obtain a multiphase quenched microstructure and achieve the components with tailored properties. There are some studies that analyze how the tool temperature affects the mechanical properties [11–13]. However, the temperature difference between the cooling tool and the heating tool is much larger when the tool is partitioned for cooling and heating, which makes the springback of the formed components more serious than conventional hot stamping. Some scholars have studied the influence of hot stamping process parameters on the springback of formed components [14–16]. They concluded that the springback of the formed component gradually decreased with the increase of blank temperature and holding time. However, for THS components with various quenching phases, the microstructure distribution has an important impact on the springback.

The springback measurement of components is a complicated process. In order to understand the effect of the quenching phase content on the component springback, a THS experiment was carried out in this paper, and the quenching phase was quantitatively measured by the color metallography and image recognition method. The springback angles of different phase contents were obtained by non-contact 3D measurement. The relationship between phase content and the springback angle was established, which can be used for the determination of the springback of THS components.

2. Experimental Study

2.1. THS Experiments

Usibor1500P boron steel with a thickness of 1 mm was used in the experiments, and the chemical composition is shown in Table 1. Figure 1 shows the THS experimental apparatus. The tool material used for this study was H13 steel and did not have any surface or hardening treatments. H13 steel has excellent comprehensive mechanical properties and high anti-tempering stability and is a common material for hot stamping tools. Cooling channels were machined inside the cooling tool to take away the heat of the blank by the water in the channels. Through installing the electric cartridge heaters controlled by the PID system, the hot tool temperatures could be varied from 25 °C and 600 °C. An air gap of 2 mm was left between the heating tool and the cooling tool to prevent heat transfer between them.

Table 1. Chemical composition of Usibor 1500P (wt.%).

C	Mn	B	Si	P	Cu	Ni	Cr	Al	Ti	Mo
0.22	1.23	0.004	0.25	0.008	0.03	0.02	0.20	0.03	0.037	<0.02

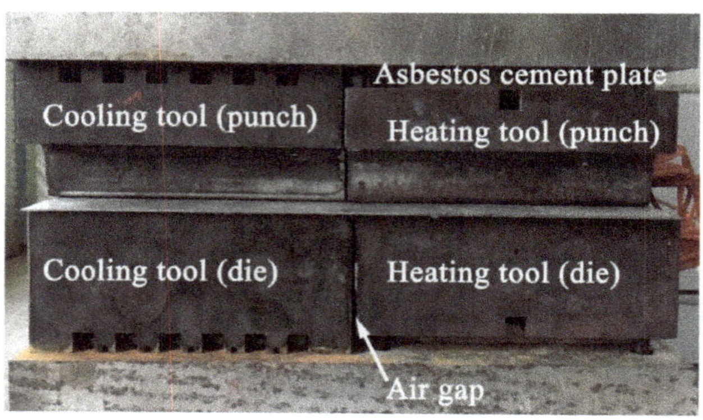

Figure 1. Schematic of the experimental apparatus.

A furnace was used to heat the blank to the temperature of 930 °C for austenizing, and the heating time was about 5 min. After the blank was austenized, it was moved to the tool to be formed and quenched. It took about 10 s to move the hot blank from the furnace to the tools. The quenching time was 10 s with the press force of 400 kN.

2.2. Metallography

Figure 2 shows the locations of microstructure observation. C represents the cold zone and H is the hot zone. To identify and measure the content of the quenched phases, the specimens were mounted in epoxy resin, ground, and polished to a mirror finish, using 400, 1000, and 2000 grit SiC paper, followed by metallographic grinding paste. Two-stage color tint etching with 4% picral solution and 10% aqueous sodium metabisulfite solution was used to observe the quenched phases. After two-stage etching, the martensite was brown, the bainite was black, and the ferrite was white with an optical microscope. Martensite, bainite, and ferrite were manually set as green, red, and blue, respectively, by ImagePro Plus 7.0 software (Media Cybernetics, Inc., Rockville, MD, USA). The area fractions of quenched phases were quantified.

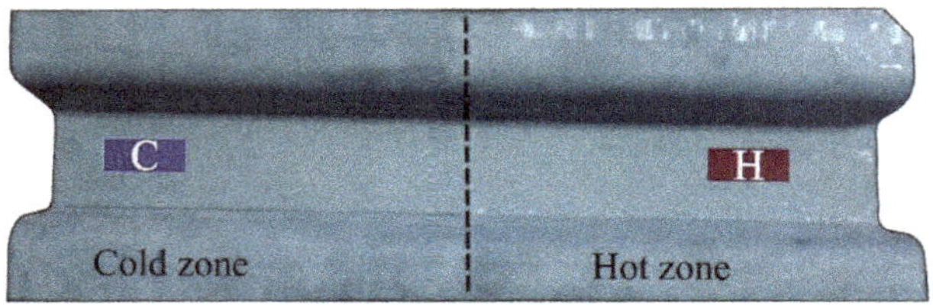

Figure 2. The cutting locations of specimens.

2.3. Springback Measurement

The springback measurement process is shown in Figure 3. The 3D scanning of the formed component was first conducted by a PRO CMM3500 optical coordinate measuring machine (NDI International, Waterloo, ON, Canada) and the point cloud data were obtained and processed by PolyWorks software (InnovMetric Software inc, Québec, Canada). By conducting calibration and alignment between polygon data and the original model of the component, the springback angle of the formed component could be accurately measured.

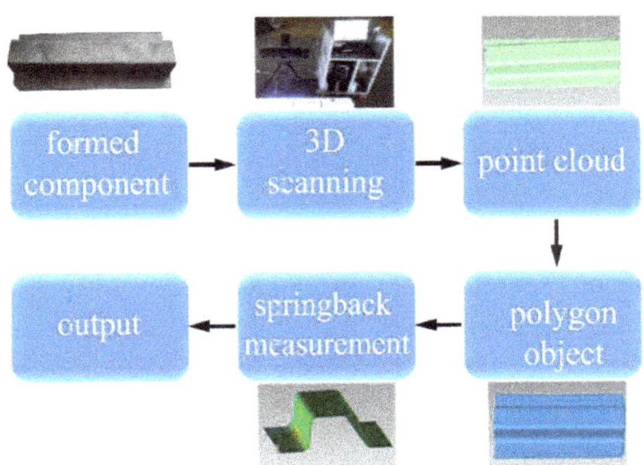

Figure 3. Flowchart of the springback measurement process.

Figure 4 shows the scheme of springback angle measurement. The outer contour and the inner contour are the cross sections of the formed component and the original component, respectively. θ_2 is the sidewall angle of the formed component and $\theta_1 = 20°$ is the original angle. $\theta_2 - \theta_1$ was defined as a springback angle.

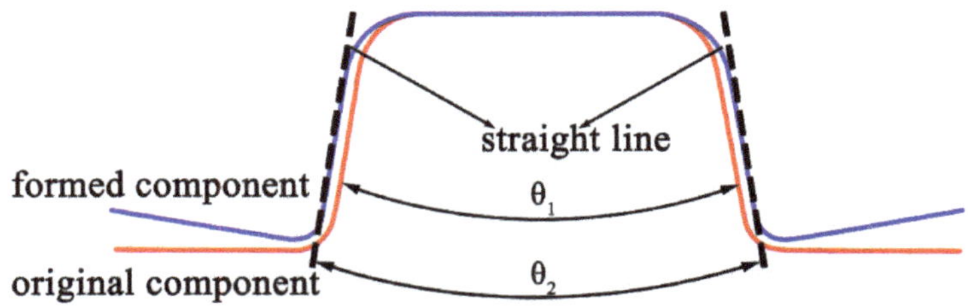

Figure 4. The schematic of the springback angle measurement.

3. Results and Discussion

3.1. Quenched Phases Analysis

Figure 5 displays color metallography of the formed component quenched at different tool temperatures and micrographs, calibrated by Image Pro Plus 7.0. It can be found that the quenched phases of the cold zone are almost full martensite at heating tool temperatures of 25 °C and 600 °C, which indicates that the heating tool temperature has little effect on the quenched phases of the cold zone. This may be because the cooling rate of the blank in the cold zone is greater than 100 °C/s, much higher than the critical speed of martensite transformation. However, the occurrence of a small amount of ferrite and bainite in the cold zone may be caused by the large plastic deformation of the U-shape component during the forming [17,18]. Martensite decreases, while ferrite and bainite increase in the hot zone with the increase of the tool temperature. When the tool temperature is higher than 200 °C, the area fraction of martensite drops dramatically and reaches 13% at the tool temperature of 600 °C. The reason may be that the start transformation temperature of martensite is 405 °C [19] and when the isothermal quenching temperature is higher than the start transformation temperature of martensite, bainite phase transition occurs during continuous cooling, resulting in the decrease of martensite. When quenched at the tool temperature from 300 °C to 600 °C, the area fraction of bainite increases with the increase of the tool temperature and is close to 70% at the tool temperature of 600 °C. The isothermal quenching for 10 s at this temperature range, which is the transition temperature range of bainite [19], results in the rapid formation and the increase of bainite. Shipway et al. [20] have also showed that bainite transition is more likely to occur than martensite transition during the isothermal quenching at this temperature range. With the increase of the tool temperature, ferrite increases slightly. George et al. [21] have demonstrated that the area fraction of ferrite is less than 10% at the tool temperature of 400 °C, which was close to the 9% ferrite in this paper. However, George et al. did not make research on the tool temperature higher than 400 °C. The studies in this paper showed that the area fraction of ferrite was 17% at the tool temperature of 600 °C.

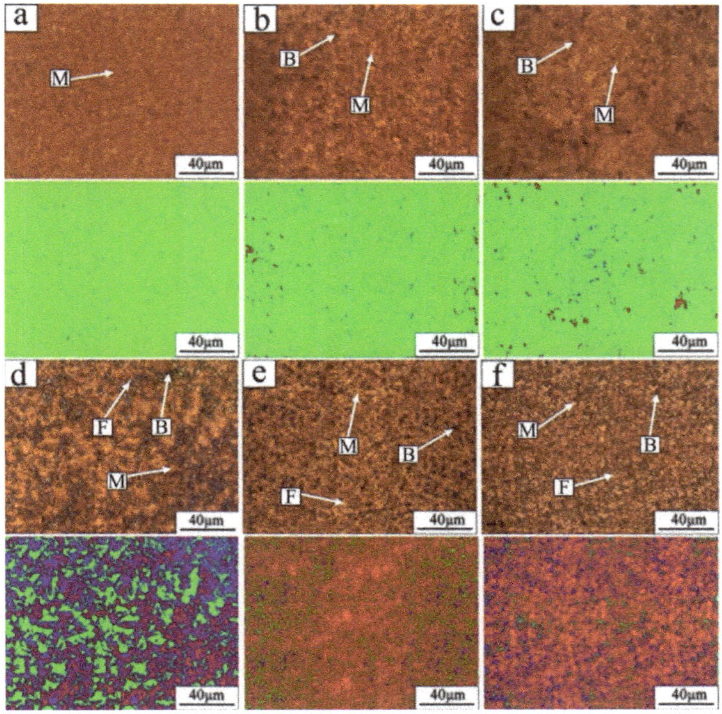

Figure 5. Two-stage color tint etched optical micrographs and manually identified microstructure images (**a**) 25 C; (**b**) 600 C; (**c**) 200 H; (**d**) 400 H; (**e**) 500 H; (**f**) 600 H. C represents the cold zone and H is the hot zone.

3.2. *Springback Results*

Springback angles under different heating tool temperatures are presented in Figure 6. No matter how the tool temperature changes in the hot zone, the spingback angles in the cold zone are almost unchanged. The reason is that the materials in the cold zone have almost the same temperature history and almost full martensite and, thus, the internal stress releases caused by phase transformation expansion and transformation plasticity are almost consistent [22]. The springback angle of the hot zone decreases with the increase of the tool temperature. When the tool temperature is higher than 300 °C, the springback angle decreases and becomes almost stable at the tool temperature higher than 550 °C. According to the relationship between the area fractions of phases and tool temperatures, it can be seen that martensite begins to decrease at the tool temperature higher than 300 °C and bainite increases dramatically. According to Åkerström and Oldenburg [23], the hardness of martensite and bainite is 510 HV and 402 HV, respectively. The decrease of the springback angle in the hot zone is due to the increase of softer bainite. When the tool temperature is higher than 550 °C, the area fractions of martensite and bainite are almost stable and, thus, there is no obvious change in the springback angle.

The relationship between the springback angle and the area fractions of martensite, bainite, and ferrite at different tool temperatures are shown in Figure 7. It can be seen that different quenched phases and their contents, caused by different tool temperatures, have great influence on the springback of the components. The springback angle shows a positive linear correlation with martensite and a negative linear correlation with bainite and ferrite. The correlation coefficient indicates that there is a noticeable linear correlation between them. With the springback angle as the dependent variable and the area fractions of martensite, bainite, and ferrite as the independent variable, a multiple linear regression analysis was carried out and the goodness of the fit, R-squared value is 0.985. The results show that there is a strong linear relationship between the springback angle and the area fractions of martensite, bainte, and ferrite. By the multi-step iterative optimization method, the relationship between the springback angle and the area fractions of quenched phases was established and is shown as Equation (1).

$$\lambda = 3.233 \times (0.0469 f_M + 0.0439 f_B + 0.0005 f_F) - 12.1286 \quad (1)$$

where λ is the springback angle and f_M, f_B, and f_F are the area fractions of martensite, bainite, and ferrite, respectively.

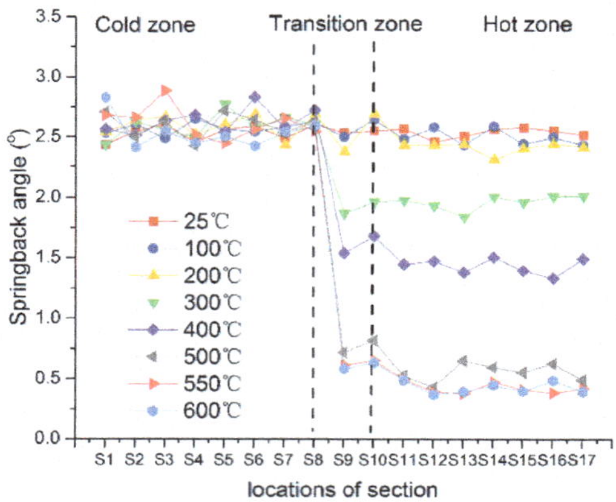

Figure 6. Springback angles under different heating tool temperatures.

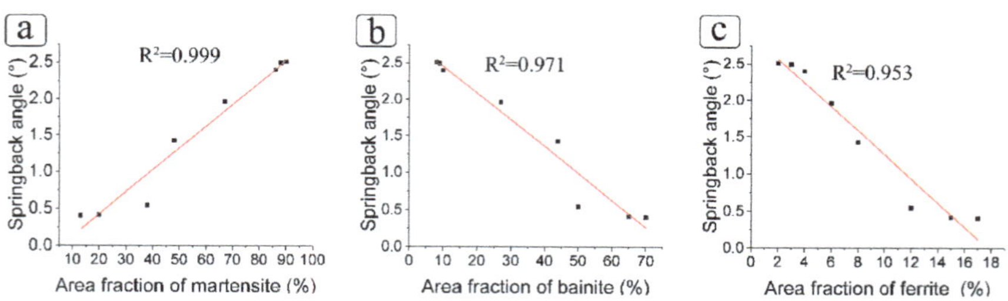

Figure 7. The relationship between the springback angle and area fractions of the quenched phases.

The springback angles calculated by Equation (1) at different heating tool temperatures are given in Table 2. It is found that the maximum relative error is 7.32% and the angle difference is 0.04°, which indicates that this equation can be used to accurately predict the springback of the formed component, based on the area fractions of the quenched phases. It is worth noting that Equation (1) is the relationship between the phase content and the springback angle, and the change of the tool geometry will lead to the change of the phase content, so after the tool geometry changes, Equation (1) is still applicable. In addition, Equation (1) is obtained in the range of ferrite content less than 20% and bainite content less than 80%, and if the phase content is beyond the range, it needs to be further verified.

Table 2. The average springback angle under different heating tool temperatures.

Heating Tool Temperature (°C)	Measured Value (°)	Calculated Value (°)
25	2.51	2.65
100	2.49	2.49
200	2.40	2.33
300	1.96	1.86
400	1.43	1.40
500	0.55	0.59
550	0.42	0.45
600	0.41	0.41

4. Conclusions

The hot stamping experiment of the U-shaped component was performed with segmented heating and a cooling tool. The relationship between the springback angle and the area fractions of martensite, bainite, and ferrite at different tool temperatures was obtained. The conclusions are as follows:

(1) The quenched microstructure of the cold zone was almost full martensite. The quenched microstructure of the hot zone changed greatly with increasing heating tool temperature. The area fraction of martensite dropped dramatically at heating tool temperatures higher than 200 °C. When the heating tool temperature was 600 °C, martensite dropped to 13% and bainite increased to 70%. Ferrite gradually increased at temperatures ranging from 25 °C to 600 °C and its maximum was close to 17%.

(2) The springback angle of the cold zone was large and remained unchanged with the increase of the heating tool temperature. When the heating tool temperature was higher than 300 °C, the springback angle of the hot zone decreased significantly. The springback angle was minimum and the change of it tended to be stable with the heating tool temperature over 550 °C.

(3) The relationship between the area fractions of quenched phases and the springback angle was established by the optimization method. A relative error analysis revealed that the relationship could be used to accurately predict the springback of a tailored formed hot stamped component.

Author Contributions: X.L. conceived and designed the experiments; X.Y. performed the experiments; Z.Z. polished the writing and grammar.

Funding: This work was supported by National Natural Science Foundation of China (51205162 and 51275203).

Conflicts of Interest: The authors declare no conflict of interest.

References

1. Mori, K.; Bariani, P.F.; Behrens, B.A.; Brosius, A.; Bruschi, S.; Maeno, T.; Merklein, M.; Yanagimoto, J. Hot stamping of ultra-high strength steel parts. *Cirp. Ann-Manuf. Technol.* **2017**, *66*, 755–777. [CrossRef]
2. Karbasian, H.; Tekkaya, A.E. A review on hot stamping. *J. Mater. Process. Technol.* **2010**, *210*, 2103–2118. [CrossRef]
3. Ying, L.; Zhao, X.; Dai, M.H.; Zhang, S.Z.; Hu, P. Crashworthiness design of quenched boron steel thin-walled structures with functionally graded strength. *Int. J. Impact Eng.* **2016**, *95*, 72–88. [CrossRef]
4. Merklein, M.; Wieland, M.; Lechner, M.; Bruschi, S.; Ghiotti, A. Hot stamping of boron steel sheets with tailored properties. *J. Mater. Process Technol.* **2016**, *228*, 11–24. [CrossRef]
5. Omer, K.; George, R.; Bardelcik, A.; Worswick, M.; Malcolm, S.; Detwiler, D. Development of a hot stamped channel section with axially tailored properties–experiments and models. *Int. J. Mater.* **2018**, *11*, 149–164. [CrossRef]
6. Vrolijk, M.; Lorenz, D.; Porzner, H.; Holecek, M. Supporting lightweight design: Virtual modeling of hot stamping with tailored properties and warm and hot formed aluminium. *Procedia Eng.* **2017**, *183*, 336–342. [CrossRef]

7. Chokshi, P.; Dashwood, R.; Hughes, D.J. Artificial neural network (ANN) based microstructural prediction model for 22MnB5 boron steel during tailored hot stamping. *Comput. Struct.* **2017**, *190*, 162–172. [CrossRef]
8. Mu, Y.; Wang, B.; Zhou, J.; Huang, X.; Li, X. Hot stamping of boron steel using partition heating for tailored properties: Experimental trials and numerical analysis. *Metall. Mater. Trans. A* **2017**, *48*, 5467–5479. [CrossRef]
9. Bardelcik, A.; Worswick, M.J.; Winkler, S.; Wells, M.A. A strain rate sensitive constitutive model for quenched boron steel with tailored properties. *Int. J. Impact Eng.* **2012**, *50*, 49–62. [CrossRef]
10. Li, F.F.; Fu, M.W.; Lin, J.P. Effect of cooling path on the phase transformation of boron steel 22mnb5 in hot stamping process. *Int. J. Adv. Manuf. Technol.* **2015**, *81*, 1391–1402. [CrossRef]
11. Omer, K.; Kortenaar, L.; Butcher, C.; Worswick, M.; Malcolm, S.; Detwiler, D. Testing of a hot stamped axial crush member with tailored properties experiments and models. *Int. J. Impact Eng.* **2017**, *103*, 12–28. [CrossRef]
12. Zhou, M.B.; Tang, J.L.; Yang, J.; Wang, C.Y. Tailored properties of a novelly designed press hardened 22MnMoB steel. *J. Iron Steel Res. Int.* **2017**, *24*, 508–512. [CrossRef]
13. Mu, Y.H.; Wang, B.Y.; Zhou, J.; Kang, Y.; Li, X.T. Heating parameters optimization of hot stamping by partition heating for tailored properties. *ISIJ. Int.* **2017**, *57*, 1442–1450. [CrossRef]
14. Nakagawa, Y.; Mori, K.I.; Maeno, T. Springback-free mechanism in hot stamping of ultra-high-strength steel parts and deformation behaviour and quenchability for thin sheet. *Int. J. Adv. Manuf. Techol.* **2018**, *95*, 459–467. [CrossRef]
15. Krinninger, M.; Opritescu, D.; Golle, R.; Volk, W. Experimental investigation of the influence of punch velocity on the springback behavior and the flat length in free bending. *Procedia CIRP* **2016**, *41*, 1066–1071. [CrossRef]
16. Bao, J.; Liu, H.; Xing, Z.; Song, B.; Yang, Y. Springback of hot stamping and die quenching with ultra-high-strength boron steel. *Eng. Rev.* **2013**, *33*, 151–156.
17. Barcellona, A.; Palmeri, D.; Worswick, M. Hot forming of boron steels using heated and cooled tooling for tailored properties. *Metall. Mater. Trans. A* **2009**, *40*, 1160–1174. [CrossRef]
18. Merklein, M.; Svec, T. Effect of plastic hot deformation on the hardness and continuous cooling transformations of 22MnB5 microalloyed boron steel. In Proceedings of the 29th IDDRG International Conference, Graz, Austria, 31 May–2 June 2010.
19. Nikravesh, M.; Naderi, M.; Akbari, G.H. Transformation kinetics of the hot stamping steel 22MnB5 in dependency of the applied deformation on the austenitic microstructure. *Mater. Sci. Eng. A Struct.* **2012**, *540*, 24–29. [CrossRef]
20. Shipway, P.H.; Bhadeshia, H.K. The effect of small stresses on the kinetics of the bainite transformation. *Mater. Sci. Eng. A Struct.* **1995**, *201*, 143–149. [CrossRef]
21. George, R.; Bardelcik, A.; Worswick, M.J. Hot forming of boron steels using heated and cooled tooling for tailored properties. *J. Mater. Process Technol.* **2012**, *212*, 2386–2399. [CrossRef]
22. Bardelcik, A.; Salisbury, C.P.; Winkler, S.; Wells, M.A.; Worswick, M.J. Effect of cooling rate on the high strain rate properties of boron steel. *Int. J. Impact Eng.* **2010**, *37*, 694–702. [CrossRef]
23. Åkerström, P.; Oldenburg, M. Austenite decomposition during press hardening of a boron steel-computer simulation and test. *J. Mater. Process Technol.* **2006**, *174*, 399–406. [CrossRef]

metals

MDPI

Article

Hot Workability of 300M Steel Investigated by In Situ and Ex Situ Compression Tests

Rongchuang Chen [1,*], Haifeng Xiao [1], Min Wang [1,*] and Jianjun Li [2]

[1] School of Materials Science and Engineering, Hubei University of Automotive Technology, Shiyan 442002, China

[2] School of Materials Science and Engineering, State Key Laboratory of Materials Processing and Die & Mould Technology, Huazhong University of Science and Technology, Wuhan 430074, China

* Correspondence: crc@hust.edu.cn (R.C.); minwang@126.com (M.W.); Tel.: +86-134-7628-4413 (R.C.)

Received: 15 July 2019; Accepted: 8 August 2019; Published: 10 August 2019

Abstract: In this work, hot compression experiments of 300M steel were performed at 900–1150 °C and 0.01–10 s^{-1}. The relation of flow stress and microstructure evolution was analyzed. The intriguing finding was that at a lower strain rate (0.01 s^{-1}), the flow stress curves were single-peaked, while at a higher strain rate (10 s^{-1}), no peak occurred. Metallographic observation results revealed the phenomenon was because dynamic recrystallization was more complete at a lower strain rate. In situ compression tests were carried out to compare with the results by ex situ compression tests. Hot working maps representing the influences of strains, strain rates, and temperatures were established. It was found that the power dissipation coefficient was not only related to the recrystallized grain size but was also related to the volume fraction of recrystallized grains. The optimal hot working parameters were suggested. This work provides comprehensive understanding of the hot workability of 300M steel in thermal compression.

Keywords: 300M steel; hot processing map; thermal compression; microstructure evolution; in situ experiments

1. Introduction

The 300M steel, an ultra-high strength steel (the ultimate tensile strength ≥1800 MPa) which is a modification of AISI 4340 steel, is often used in high performance auto parts, aircraft landing gear, airframe parts, and many other high strength applications because of the outstanding mechanical properties. To ensure the best service performance, these parts usually undergo hot forging. The selection of hot working parameters has been a major concern in isothermal compression of 300M steel, because it not only affects the microstructure at high temperatures, but also affects the microstructures after cooling and service performance. However, the hot workability of the 300M steel was not well understood, resulting in grain coarsening or even cracking in practical forging of large parts [1–3]. Therefore, a comprehensive understanding of the hot workability of 300M steel in thermal compression is still needed.

The hot working maps of 300M steel or similar steels have been investigated by many researchers in recent years. A physically based flow stress model and the processing maps of a medium-carbon low-alloy steel, 34CrNiMo6, was proposed by Gong et al. [4], and it was pointed out that two optimal processing parameter domains existed, one at 1050–1130 °C and 0.005–0.03 s^{-1}, and the other at 1130–1200 °C and 0.03–0.36 s^{-1}. The constitutive model and processing maps of 4340 steel were constructed by Sanrutsadakorn et al. [5], and an optimal hot working condition was suggested at 1050–1200 °C and 0.01–0.1 s^{-1}. Also, for 4340 steel, the hot working maps were established by Łukaszek-Sołek et al. [6], and explained via microstructure observations, but a different workability parameter, at 1050–1200 °C and 3–57 s^{-1}, was suggested. The processing maps of 300M steel were formed by Luo et al. [7], and an optimal hot working parameter range was recommended at 1100–1140 °C and 0.1–0.16 s^{-1} via microstructure observations. Based on a constitutive model coupling average grain size, a processing map of 300M steel was calibrated by Sun et al. [8], and a broader hot working parameter range was determined at 900–1140 °C and 0.1–1.91 s^{-1}. These researches have laid a solid foundation for the present investigation.

Up to date, it is generally accepted that the grain evolution is the intrinsic reason for flow stress evolution [9], and the grain morphology is able to interpret the deformation mechanisms of different domains in the processing maps. The microstructure evolution of 300M steel has been studied by Chen et al. [10–13], Liu et al. [14], and Zeng et al. [15]. However, the influence of the dynamic recrystallization volume fraction on the processing map of 300M steel has not been systematically investigated. Besides, the optimal strain rate should not be too narrow to avoid difficulty in practical application [7], and an undesirable mixed grain defect was found under current suggested hot working conditions [8]. Therefore, problems about the processing map of 300M steel still exist, and systematic research on the microstructure evolution and hot workability in thermal compression of 300M steel is in need.

Accordingly, in situ and ex situ compression tests will be carried out. The hot working maps will be established. Based on the analysis of the experiment results, the optimal hot working parameters will be suggested. This work provides comprehensive understanding of the hot workability of 300M steel in thermal compression.

2. Materials and Experiments

2.1. Materials

The as-received material was annealed. The diameter of the as-received 300M steel rod was 300 mm, and height 1000 mm. The chemical composition was quantitatively characterized by X-ray fluorescence (XRF) as 0.39C-0.808Mn-0.086V-0.824Ni-0.435Mo-2.562Si-0.896Cr-0.017S with balanced Fe (weight %).

2.2. Hot Compression Tests

Traditional ex situ metallographic observation was implemented to investigate the microstructure evolution of 300M steel in thermal compression. Specimens for the hot compression tests were wire-electrode cut from the ingot, and they were turned into cylinders whose diameter was 8 mm, the height 12 mm. The hot compression tests were performed on a Gleeble 3500 machine (Dynamic Systems Inc., Austin, TX, USA). The specimens were heated at 3.3 °C/s to 1200 °C. After holding for 240 s, the specimens were cooled to deformation temperatures, compressed at different strain rates to the strain of 0.95, and quenched thereafter. The compression temperatures were 900 °C, 950 °C, 1000 °C, 1050 °C, 1100 °C, and 1150 °C, which were chosen to cover the typical forging temperature range of this material [16]. The selection of strain rates (0.01, 0.1, 1, and 10 s^{-1}) has considered the usual strain rate range in the die forging process [17]. The temperatures were measured via a thermal-couple welded on the half height of the specimen surface, and automatically controlled during heating, holding, compression, and cooling by a computer to obtain specified temperatures. Punch positions were also automatically monitored to get constant strain rates. The true stress was calculated by the punch force and instant cross-section area of the specimen. The logarithmic strain was calculated according to the height reduction of the specimen. Friction was reduced by placing tantalum sheets between the punch and specimen end. The stress–strain curves underwent noise reduction.

The microstructures of deformed specimens were characterized by metallographic observations on an optical microscope (VHX-1000C, Keyence Co., Osaka, Japan). The specimens were cut via the symmetry faces which passed the axis. Specimens were grinded, polished, and etched using etchant of saturated picrate, carbon tetrachloride, and concentrated hydrochloric acid with a volume ratio of

6:1:1:3. The temperature for etching was ~40 °C. The middle regions of the specimens were observed. At least five photos with square areas of 0.4 mm × 0.4 mm were taken for each specimen.

2.3. In Situ Compression Tests

To investigate continuously the influence of strain on microstructure evolution of 300M steel, in situ compression tests were performed on a confocal laser scanning microscope (CLSM, VL2000DX-SVF15FTC, Yonekura MFG Co, Osaka, Japan). The experimental setup is schematically shown in Figure 1. A square area of 6 mm × 6 mm on the dumbbell-shaped specimen surface was mechanically polished for in situ observation. The specimen was calibrated to avoid compression instability. The marking length was 15 mm, and the cross-section shape was a rectangle with the size of 4 mm × 6 mm [18], the design of which has considered the installation room size and punch force limitations. The test specimens were gripped at both ends by screw threads on the punches. During the in situ compression test, the microstructure evolutions of the specimen could be visualized on the polished surface because of the volatilization of alloy elements at grain boundaries. Further technique details can be referred to in the literature [11]. The compression temperature, strain rate, and strain were 1100 °C, 0.01 s^{-1}, and 0.9, respectively.

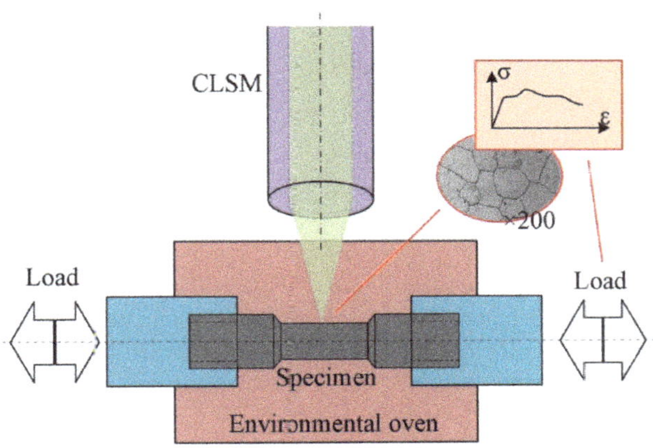

Figure 1. Illustration of the experimental setup.

3. Results and Discussion

3.1. Flow Behavior

The flow stress curves of 300M steel are shown in Figure 2. The contradicting effect of work hardening and dynamic recovery softening led to an increase of flow stress in the initial stage of compression, after which the dynamic recrystallization was triggered because the dislocation density reached a critical value, and an intensive softening due to dynamic recrystallization took place, resulting in a gradual drop of flow stress near the strain of ~0.3. When the dynamic recrystallization was completed, the work hardening and softening were balanced, and the flow stresses were almost stable. It should be noted that different flow behaviors were shown at different strain rates. At 0.01 s^{-1}, as shown in Figure 2a, the flow stress curves were single-peaked, while at 10 s^{-1}, as shown in Figure 2d, no peak occurred. This was probably because dynamic recrystallization was more complete at a lower strain rate, so the dynamic recrystallization softening was more intensive, thereby resulting in a typical single-peaked discontinuous dynamic recrystallization flow stress curve according to Sakai et al. [9]. Further metallographic observations were carried out in Section 3.2 to confirm the metallurgy reason.

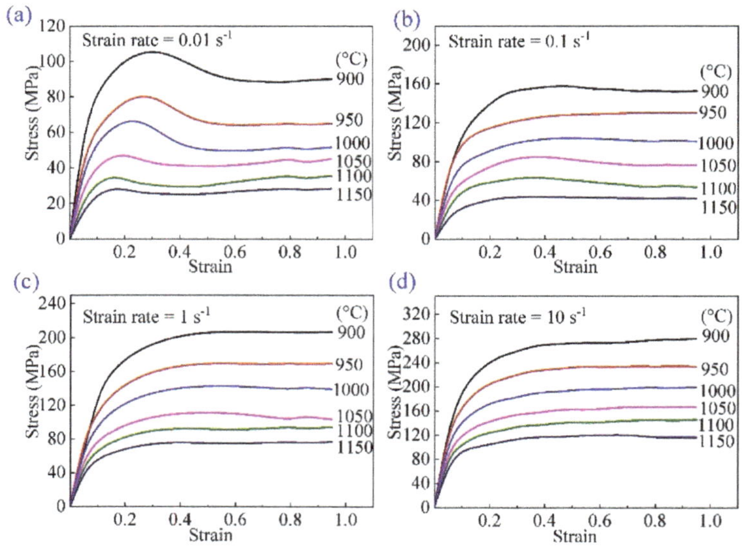

Figure 2. Flow stress curves of 300M steel compressed at (**a**) 0.01 s^{-1}, (**b**) 0.1 s^{-1}, (**c**) 1 s^{-1}, and (**d**) 10 s^{-1}.

3.2. Microstructure Evolution

In order to investigate the effect of temperature on microstructure evolutions of 300M steel, the microstructures of deformed specimens at various temperatures (900–1150 °C), at two strain rates (0.01 s^{-1}, 10 s^{-1}), and at the strain of 0.95 were observed on the optical microscope, shown in Figures 3 and 4. At 0.01 s^{-1}, a great number of fine grains nucleated at curved boundaries of the coarse initial grains at 900 °C, as shown in Figure 3a, because the high dislocation densities at the grain boundaries were beneficial for recrystallization. However, the dislocation density inside the initial grains was comparatively lower, resulting in incomplete recrystallization microstructures. The coarse initial grains disappeared when the temperature increased to 1000 °C, as shown in Figure 3b,c, due to full recrystallization. Moreover, grains were gradually coarsened with increasing temperature, as shown in Figure 3a–f, which were also shown in the literature [10,14,15]. Under lower temperatures and higher strain rates, grain coarsening was inhibited, and the effect of grain refinement caused by dynamic recrystallization was more obvious. However, under higher temperatures and lower strain rates, the recrystallized grains grow rapidly once generated, and grain coarsening due to grain boundary curvature was dominant, resulting in coarse grains.

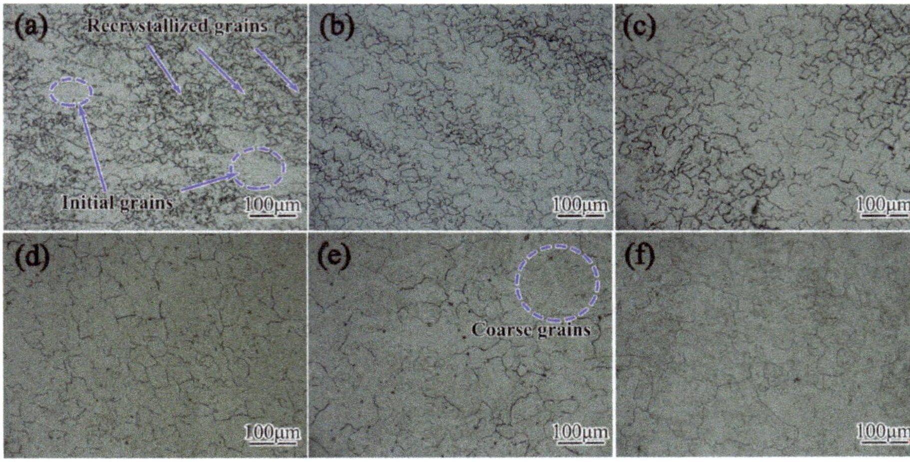

Figure 3. Microstructures of 300M steel compressed at 0.01 s^{-1} and at (**a**) 900 °C, (**b**) 950 °C, (**c**) 1000 °C, (**d**) 1050 °C, (**e**) 1100 °C, and (**f**) 1150 °C.

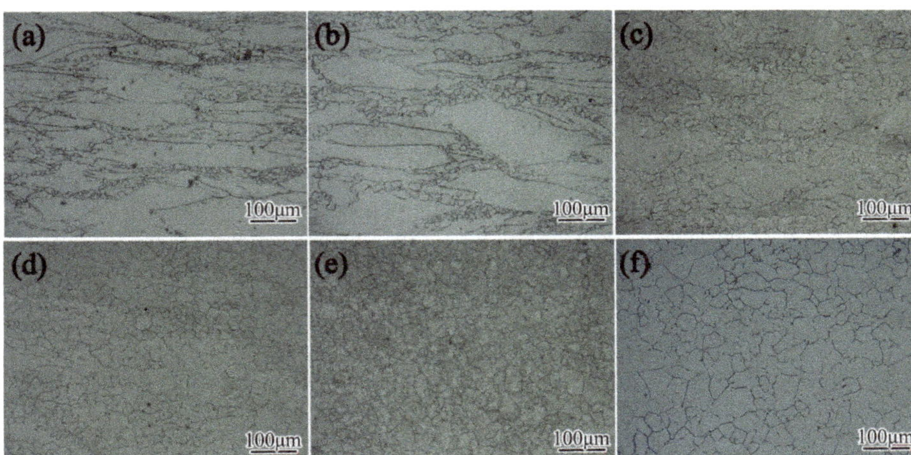

Figure 4. Microstructures of 300M steel compressed at 10 s^{-1} and at (**a**) 900 °C, (**b**) 950 °C, (**c**) 1000 °C, (**d**) 1050 °C, (**e**) 1100 °C, and (**f**) 1150 °C.

The microstructures of 300M steel compressed at 10 s^{-1} are shown in Figure 4. A mixed microstructure of fine recrystallized grains and coarse initial grains was obtained at 900–1000 °C, as shown in Figure 4a–c. With increasing temperature, a full recrystallized microstructure was formed at 1050–1150 °C, as shown in Figure 4d–f, and with increasing temperature, the average grain size increased. The mean grain boundary migration rate was less than 0.2 μm/s at 900–1150 °C according to a previous study [12], and under the strain rate of 10 s^{-1}, the average grain size grew by 0.02 μm at most during the compression process, which was negligible compared with the average grain size (≥5 μm). Thus, the effect of grain coarsening due to grain boundary curvature could be neglected. The newly generated recrystallization grain size was not only temperature dependent, but also strain rate dependent. The lower the strain rate, the longer it needed to reach the same strain and therefore the coarser the recrystallized grains. Moreover, it can be seen in Figure 4 that with the increase of temperature, the volume fraction of recrystallized grains increased gradually. That was due to the fact that at lower temperatures the deformation stored energy in the material was insufficient to overcome the energy barrier for recrystallization, and at higher temperatures, the recrystallization more easily occurred, resulting in a higher recrystallization volume fraction.

To investigate the influence of strain rate, the microstructures of deformed specimens at various strain rates (0.01, 0.1, 1, and 10 s^{-1}), at two temperatures (900 and 1150 °C), and at the strain of 0.95 were observed on the optical microscope, shown in Figures 5 and 6. At 900 °C, as shown in Figure 5, coarse initial grains were squashed, around which many fine recrystallized grains emerged, thus obtaining a mixture of fine recrystallized grains and coarse initial grains. It can also be seen that with increasing strain rate, the percentage of recrystallized grains and the average grain size decreased. This was because the nuclei formed near the grain boundaries due to high dislocation density, and when the strain rate was low, as shown in Figure 5a, the nuclei had enough time to grow, thus consuming the coarse initial grains, resulting in a higher volume fraction of the recrystallized grains and a larger average grain size. However, when the strain rate was high, as shown in Figure 5b–d, the nuclei did not have enough time to grow, and the nuclei continuously formed near grain boundaries, leading to the "package" of many small nuclei near grain boundaries.

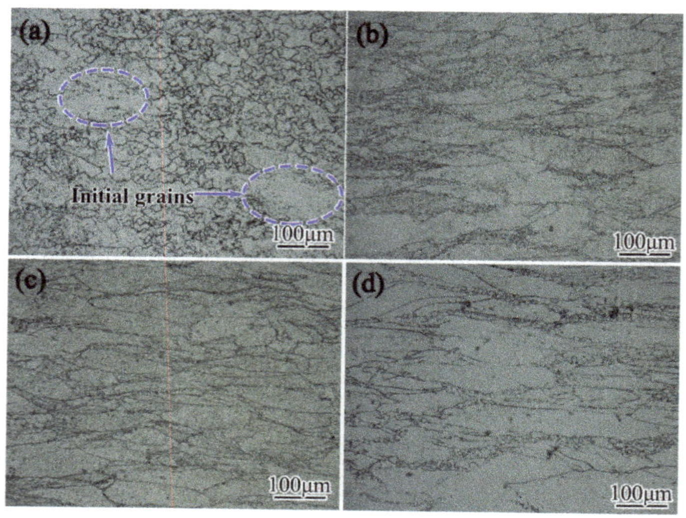

Figure 5. Microstructures of 300M steel compressed at 900 °C and at (**a**) 0.01 s^{-1}, (**b**) 0.1 s^{-1}, (**c**) 1 s^{-1}, and (**d**) 10 s^{-1}.

The microstructures of 300M steel compressed at 900 °C are shown in Figure 6. At the temperature of 1150 °C, the recrystallization was completed. With increasing strain rate, the grain size gradually became smaller, because the recrystallized grains had less time to grow, which was also found in the literature [19–21]. The results in Figures 5 and 6 could explain the different flow behaviors of 300M steel at 0.01 s^{-1} and 10 s^{-1} in Figure 2. At 10 s^{-1}, the overall volume fraction of recrystallized grains was not high. Dynamic recovery dominated, and therefore, it showed monotonically increasing type stress–strain curves. While at 0.01 s^{-1}, the dynamic recrystallization was completed, as shown in Figure 6, and dynamic recrystallization dominated. Thus, it showed single-peaked type stress–strain curves. The dynamic recrystallization kinetic model and the average grain size model have been established by the authors previously [10]. The incomplete recrystallization was due to insufficient thermal activation at lower temperatures. By increasing the deformation temperature, reducing the strain rate, or increasing the strain, the recrystallization volume fraction can be increased.

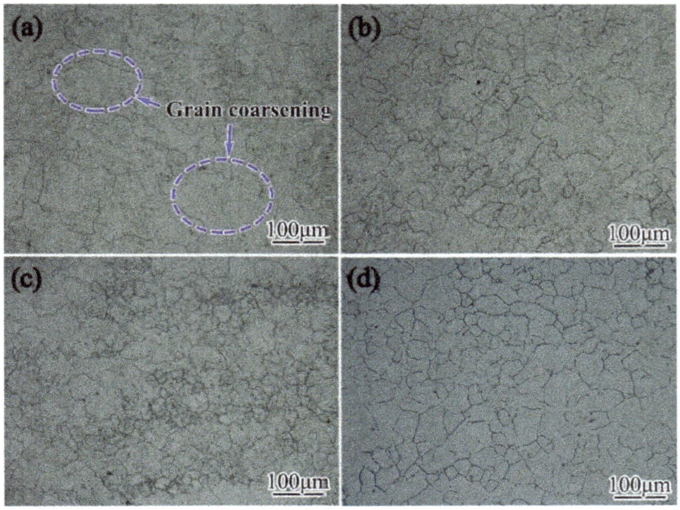

Figure 6. Microstructures of 300M steel compressed at 1150 °C and at (**a**) 0.01 s^{-1}, (**b**) 0.1 s^{-1}, (**c**) 1 s^{-1}, and (**d**) 10 s^{-1}.

Since almost all austenite grains transformed to small martensite laths during quenching, reconstructing the orientation relationship of austenite before quenching by ex situ characterizations on electron backscattered diffraction (EBSD) or transmission electron microscope (TEM) is very hard, although the orientation relationship of austenite and transformed martensite laths has been studied in several research works [22–24]. Traditional ex situ metallographic observation on the optical microscope, or in situ observation on the high temperature confocal laser scanning microscope, is still the most effective way to study the microstructure evolutions of 300M steel during compression. To compare with the results of traditional metallographic observation on the optical microscope, in situ compression tests were carried out at 1100 °C and 0.01 s^{-1}. The photos taken by the microscope at the strain of 0, 0.1, 0.2, 0.3, 0.4, 0.5, and 0.9 are shown in Figure 7a–g. It can be seen in Figure 7a that equiaxed grains were obtained after four minutes' holding at 1200 °C. When the deformation started, the dark areas at grain boundaries became larger and larger, as shown in Figure 7b–g. That was because the observed surface of the specimen became rough, resulting in a decreased light reflectivity. The merged graph of Figure 7a–f is shown in Figure 7h to represent the microstructures at various strains. With increasing strain, the grains became finer and finer due to the crush of initial grains and dynamic recrystallization. Although the in situ observation results in this research only provide qualitative information about the recrystallized grain size or recrystallization kinetics, it is still very important to observe the grain evolution process of 300M steel in thermal compression, and this is one of the few attempts that have succeeded in observing the grain evolution of steel during high temperature compression in recent years.

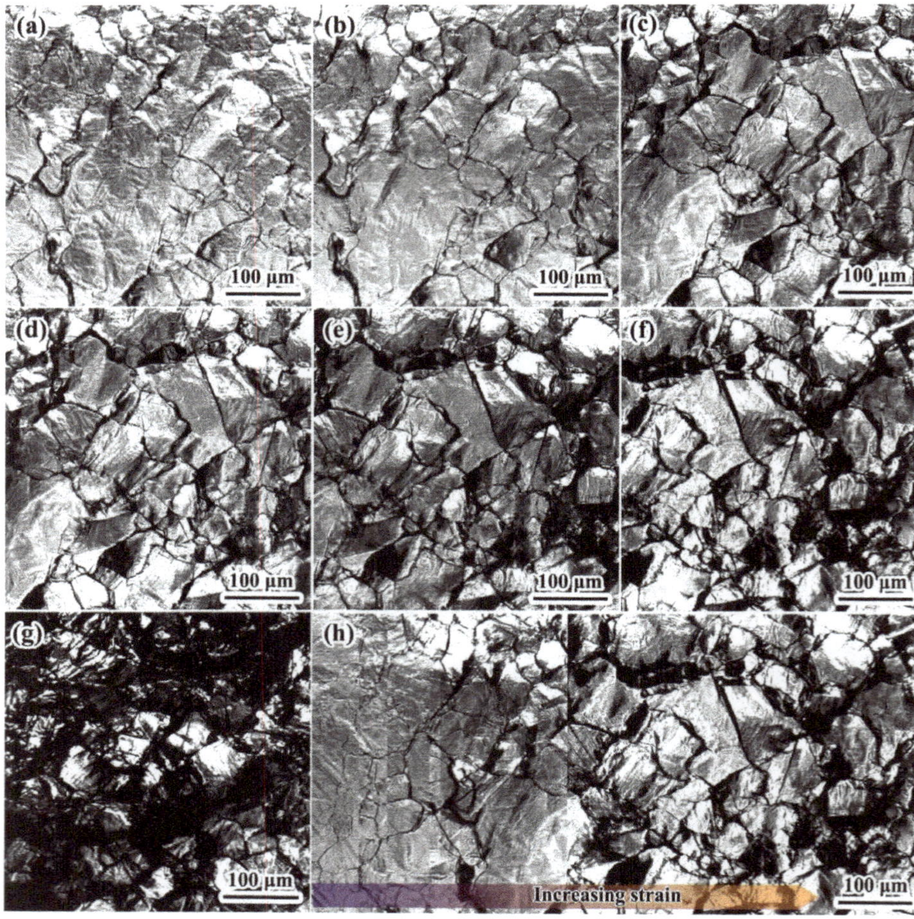

Figure 7. In situ observation results of the microstructure evolutions of 300M steel compressed at the temperature of 1100 °C, at the strain rate of 0.01 s^{-1}, and at the strain of (**a**) 0.1, (**b**) 0.2, (**c**) 0.3, (**d**) 0.4, (**e**) 0.5, (**f**) 0.6, and (**g**) 0.9. The merged graph of (**a–f**) was shown in (**h**) to represent the microstructures at various strains.

3.3. Hot Working Map

The metallographic results showed that under higher temperatures or lower strain rates, recrystallization was more likely to complete, which was very beneficial to obtain uniform microstructures, but meanwhile, grain coarsening was also likely to more easily to occur, which was undesirable for industrial applications. Thus, a thermal processing map may facilitate the selection of process parameters to obtain uniform, small recrystallization microstructures. Previous studies have focused only on the relationship between average grain size and thermal processing map, but the influence of the dynamic recrystallization volume fraction on processing maps of 300M steel has not been systematically investigated. Besides, the optimal strain rate should not be too narrow to avoid difficulty in practical application [7], and an undesirable mixed grain defect was found under current suggested hot working conditions [8], as shown in Figure 5b,c. Therefore, the processing maps at different temperatures, strain rates, and strains were established. The influences of average grain size and dynamic recrystallization volume fraction were systematically investigated.

Seeing forgings as isolated systems whose input energy was the deformation work, and output energy was the consumption by temperature rising and microstructure evolution, the input energy (P) was calculated by [25]

$$P = G + J = \int_0^{\dot{\varepsilon}} \sigma \, d\dot{\varepsilon} + \int_0^{\sigma} \dot{\varepsilon} d\sigma \tag{1}$$

where G was the energy consumption by temperature rising, and J by microstructure evolution. The variables of σ, ε, and $\dot{\varepsilon}$ denoted stress (MPa), strain (1), and strain rate (s^{-1}), respectively. The power dissipation coefficient (η) representative of the fraction of energy consumption by microstructure evolution was expressed as

$$\eta = \frac{2\frac{\partial ln\sigma}{\partial ln\dot{\varepsilon}}}{\frac{\partial ln\sigma}{\partial ln\dot{\varepsilon}} + 1}. \tag{2}$$

The instability coefficient (ζ) evaluating the deformation instability was calculated by

$$\zeta = \frac{\partial \lg(\eta/2)}{\partial \log(\dot{\varepsilon})} + \frac{\partial ln\sigma}{\partial ln\dot{\varepsilon}}. \tag{3}$$

The stresses at various strains could be obtained via thermal compressions, and the processing maps of 300M steel were established according to Equations (2) and (3) and are shown in Figure 8. The processing maps at various strains are shown in Figure 9. The contours show the value of the power dissipation coefficient, and the shaded areas show the parameters under which deformation instability occurred. It can be seen that two deformation instability domains existed. One located at 0.03–0.7 s^{-1}, 900–1100 °C, and 0.2–0.8 (strain). The other located at 3.2–10 s^{-1}, 1100–1150 °C, and 0.2–0.8. Representative microstructure of the two instability domains are shown in Figures 5b and 6a. It can be seen that incomplete recrystallization and grain coarsening occurred, respectively. The average dissipation coefficient of these two instability domains were in the range of 15–25%, indicating that a comparatively large part of deformation work was converted to heat, while a small part was consumed by microstructure evolution. It can also be seen from Figures 8 and 9 that two high dissipation coefficient domains existed. One located at 0.1–3 s^{-1}, 1100–1150 °C, and 0.4–0.8. The other located at 0.01–0.06 s^{-1}, 900–1100 °C, and 0.2–0.6. Representative microstructure of the two high dissipation coefficient domains are shown in Figures 6c and 3c. It can be seen from the microstructure photos that recrystallization was completed. The average dissipation coefficient of these two high coefficient domains were in the range of 35–45%, which was ideal for a uniform recrystallization microstructure.

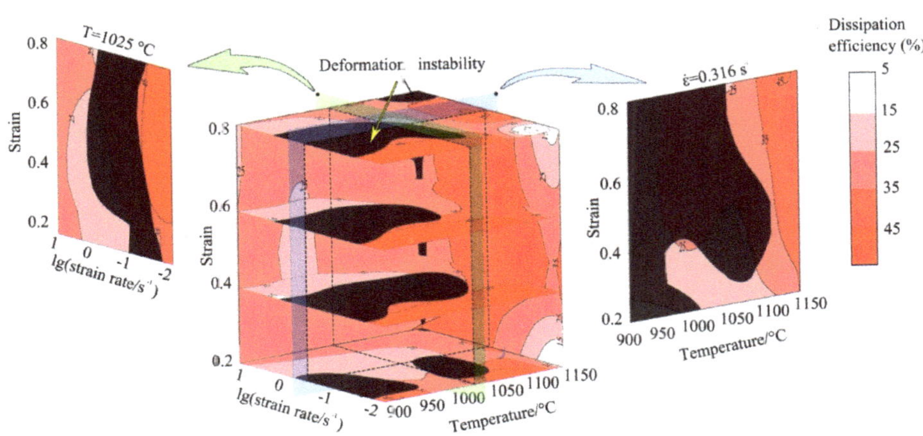

Figure 8. The three dimensional hot working maps of 300M steel.

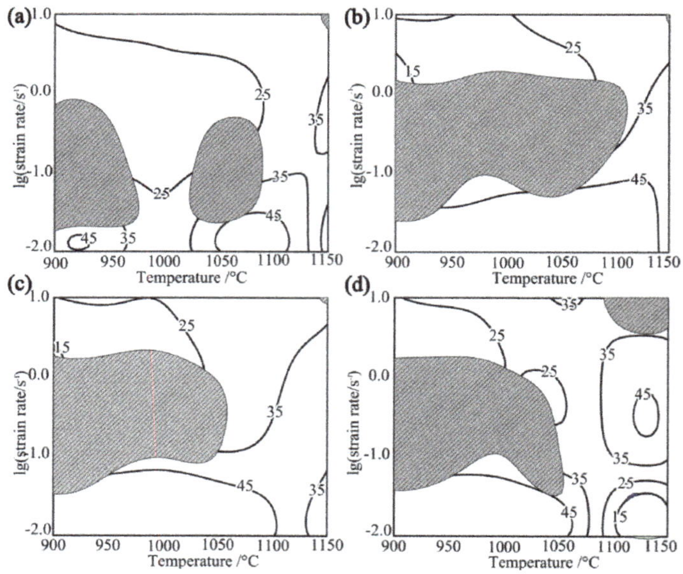

Figure 9. Hot working maps of 300M steel at the strain of (**a**) 0.2, (**b**) 0.4, (**c**) 0.6, and (**d**) 0.8.

Processing maps at specific temperatures and strain rates could be obtained by slicing and interpolating the 3D processing maps at specific temperatures and strain rates, shown in Figures 10 and 11. It can be seen from Figure 10a that at 0.01 s^{-1}, a stable deformation domain with the dissipation coefficient of 35–45% could be obtained. Typical microstructures of this domain are shown in Figure 3a–f. However, the instability domain occupied most parts of the hot working maps at 900–1100 °C when the strain rate increased from 0.1 to 1 s^{-1}, as shown in Figure 10b,c, indicating that incomplete recrystallization and localized material flow may be the reason. The instability domain at 900–1100 °C disappeared when the strain rate further increased to 10 s^{-1}, as shown in Figure 10d, and the dissipation coefficient was around 25%, meaning that incomplete dynamic recrystallization took place, as shown in Figure 4a–e. Metallographic observation results in Figure 5b,c proved that incomplete recrystallization and mixed grain defect occurred. Thus, for the 300M steel, it is easier to obtain a full recrystallization microstructure at 0.01 s^{-1}, while at 10 s^{-1} incomplete recrystallization was obtained. At a moderate strain rate (0.1–1 s^{-1}), the deformation temperature should be above 1100 °C.

The processing maps at various temperatures are shown in Figure 11. As shown in Figure 11a, a deformation instability domain at 0.016–1.3 s^{-1} was found at 900 °C. Representative microstructures are shown in Figure 5a–c. It could be explained that at a lower temperature (~900°C), it was difficult to overcome the energy barrier of dynamic recrystallization, and the material mainly underwent incomplete recrystallization and localized flow, so most energy was converted to heat, resulting in a low dissipation coefficient of ~15%. It is shown in Figure 11b that the deformation instability domain was separated into two parts. One part located at the strain rate range of 0.016–0.71 s^{-1} and strain range of 0.2–0.3, and the other at 0.08–0.5 s^{-1} and 0.4–0.8. The size of deformation instability domain shrank in the strain rate range of 0.02–1 s^{-1} and strain range of 0.2–0.6 at the temperature of 1100 °C, as shown in Figure 11c.

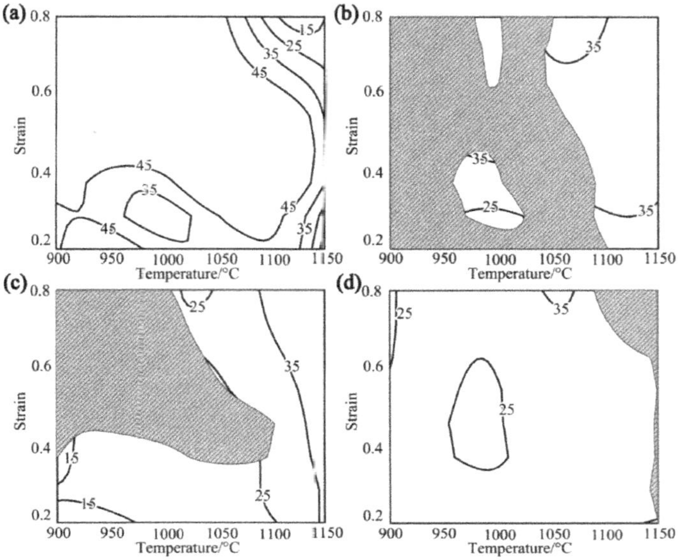

Figure 10. Hot working maps of 300M steel at the strain rate of (**a**) 0.01 s^{-1}, (**b**) 0.1 s^{-1}, (**c**) 1 s^{-1}, and (**d**) 10 s^{-1}.

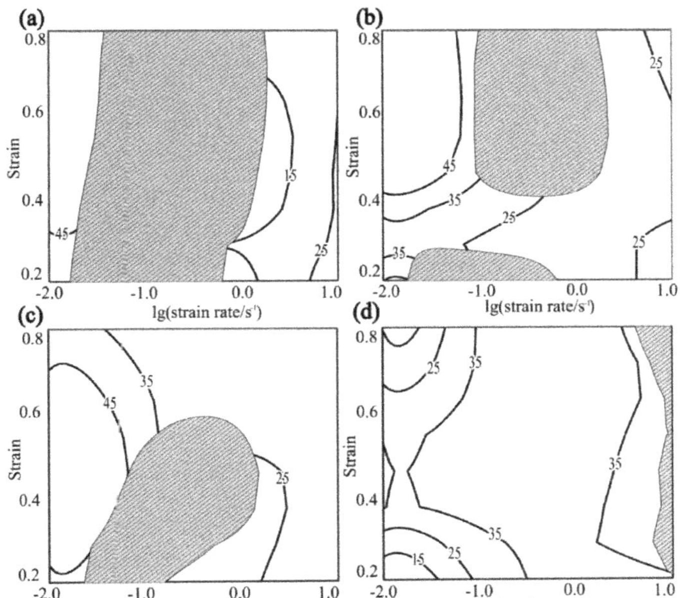

Figure 11. Hot working maps of 300M steel at the temperature of (**a**) 900 °C, (**b**) 1000 °C, (**c**) 1100 °C, and (**d**) 1150 °C.

It was worth noting that at 1000–1100 °C, as shown in Figure 11b,c, a high dissipation coefficient domain (≥35%) existed in the strain rate range of 0.01–0.063 s^{-1} and in the strain range of 0.4–0.8, indicating that an optimal microstructure (similar to Figure 11c–e) could be obtained due to dynamic recrystallization. When the deformation temperature further increased to 1150 °C, as shown in

Figure 11d that the deformation instability domain located in the high strain rates region (~10 s^{-1}), and the dissipation coefficient was around 35% in the strain rate range of 0.01–3.2 s^{-1} and strain range of 0.2–0.8. Comparing the results of Figure 11a–d, it was found that a smaller deformation instability domain was obtained at a higher deformation temperature. Therefore, in order to obtain a fine and uniform microstructure, the hot working parameters were suggested at 0.1–3 s^{-1} and 1100–1150 °C.

The suggested hot working parameters in the present research is compared with the optimal parameters of 300M steel and 4340 steel in the literature, shown in Figure 12. It is worth mentioning that the optimal strain rate was not proposed in reference [1], thus the upper and lower boundaries of the strain rate were left empty. The optimal hot working parameter range in the present research was much smaller than the existing commercial recommendation by Speich et al. [1], and it has covered the parameter range by Luo et al. [7]. Besides, the 300M steel showed a different optimal hot working parameter range from 4340 steel due to the addition of silicon and vanadium, which on one hand has increased the thermal strength of steel, on the other, decreased the hot workability. In practical production of 300M steel forgings the optimal hot working parameters could be implemented by the deformation temperature selection, punch speed selection, and pre-forging shape designing. The optimal dissipation coefficient was between 35% and 40% according to previous research [26]. When the deformation temperature and the pre-forging shape was restrained, the punch speed curves could be continuously adjusted to ensure that the average dissipation coefficient was between 35% and 40%, and under such circumstances, the finite element simulation was beneficial for the optimization of punch speed curves.

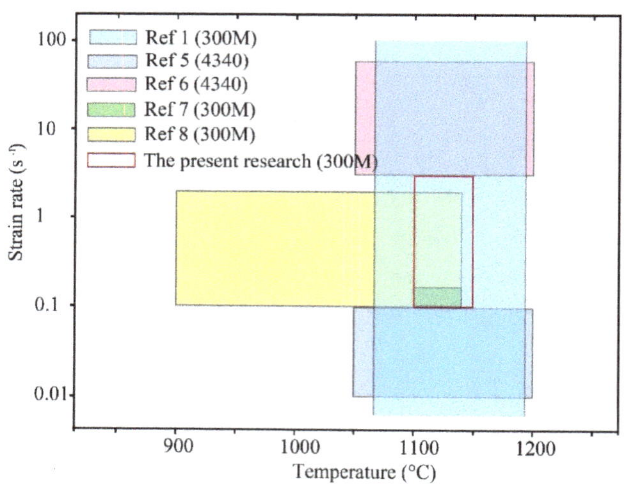

Figure 12. Comparison of the optimal hot working parameters in the present research with the optimal parameters in the literature.

4. Conclusions

The following conclusions can be drawn based on the present investigation.

(1) Different shapes of flow stress curves were obtained at different strain rates, and the shape change was explained by microstructure observations. At 10 s^{-1}, the overall volume fraction of recrystallized grains was not high. Dynamic recovery dominated, and therefore, it showed monotonically increasing dynamic recovery type stress–strain curves. While at 0.01 s^{-1}, the dynamic recrystallization was completed, and dynamic recrystallization dominated. Thus, it showed single-peaked dynamic recrystallization type stress–strain curves.

(2) In situ compression results showed with increasing strain, the grains became finer and finer due to crushing of initial grains and dynamic recrystallization. This investigation is one of the few attempts that has succeeded in observing the grain evolution of steel during high temperature compression in recent years.
(3) The hot processing maps were not only related to the recrystallized grain size, but also related to the volume fraction of recrystallized grains. The dissipation coefficients were comparatively higher at higher temperatures and slower strain rates, which was usually due to complete dynamic recrystallization and severe grain growth. While at lower temperatures and higher strain rates, the dissipation coefficients were low, incomplete recrystallization and mixed grain defect occurred.
(4) The optimal working parameters were suggested at 0.1–3 s^{-1} and 1100–1150 °C. When the deformation temperature and the pre-forging shape were restrained, the punch speed curves could be continuously adjusted to ensure that the average dissipation coefficient was between 35% and 40%.

Author Contributions: Writing-original draft preparation, R.C.; writing-review & editing, H.X.; funding acquisition, M.W. and J.L.

Funding: This study was funded by the National Natural Science Foundation of China (51435007), the Natural Science Foundation of Hubei province (2017CFB587), the State Key Laboratory of Materials Processing and Die & Mould Technology in Huazhong University of Science and Technology (P2020-015), the Outstanding Young Scientific &Technological Innovation Team Plan of Colleges and Universities in Hubei Province (T201518), the Hubei Key Laboratory of Automotive Power Train and Electronic Control in Hubei University of Automotive Technology (ZDK1201601), and the Doctoral Scientific Research Fund of Hubei University of Automotive Technology (BK201901).

Conflicts of Interest: The authors declare that they have no conflict of interest.

References

1. Speich, G.R. Properties and Selection Irons and Steels, Ultrahigh-Strength Steels section. In *Metals Handbook*; ASM International: Materials Park, OH, USA, 2005; Volume 1, pp. 1118–1170.
2. Dong, H.; Zhang, S.H.; Yong, Q.L. Steel and iron materials engineering. In *China Materials Engineering Canon*; Chemical Industry Press: Beijing, China, 2006; Volume 3, pp. 1048–1049.
3. Skubisz, P.; Sinczak, J. Properties of direct-quenched aircraft forged component made of ultrahigh-strength steel 300M. *Aircr. Eng. Aerosp. Technol.* **2018**, *90*, 713–719. [CrossRef]
4. Gong, B.; Duan, X.W.; Liu, J.S.; Liu, J.J. A physically based constitutive model of as-forged 34CrNiMo6 steel and processing maps for hot working. *Vacuum* **2018**, *155*, 345–357. [CrossRef]
5. Sanrutsadakorn, A.; Uthaisangsuk, V.; Suranuntchai, S.; Thossatheppitak, B. Investigation of hot deformation characteristics of AISI 4340 steel using processing maps. *Adv. Mater. Res.* **2013**, *683*, 301–306. [CrossRef]
6. Łukaszek-Sołek, A.; Krawczyk, J.; Sleboda, T.; Grelowski, J. Optimization of the hot forging parameters for 4340 steel by processing maps. *J. Mater. Res. Technol.* **2019**, *8*, 3281–3290. [CrossRef]
7. Luo, J.; Li, M.Q.; Liu, Y.G.; Sun, H.M. The deformation behavior in isothermal compression of 300M ultrahigh-strength steel. *Mater. Sci. Eng. A* **2012**, *534*, 314–322. [CrossRef]
8. Sun, H.M.; Li, M.Q.; Liu, Y.G. Development of processing map coupling grain size for the isothermal compression of 300M steel. *Mater. Sci. Eng. A* **2014**, *595*, 77–85. [CrossRef]
9. Sakai, T.; Belyakov, A.; Kaibyshev, R.; Miura, H.; Jonas, J.J. Dynamic and post-dynamic recrystallization under hot, cold and severe plastic deformation conditions. *Prog. Mater. Sci.* **2014**, *60*, 130–207. [CrossRef]
10. Chen, R.; Guo, P.; Zheng, Z.; Li, J.; Feng, F. Dislocation Based Flow Stress Model of 300M Steel in Isothermal Compression Process. *Materials* **2018**, *11*, 972. [CrossRef]
11. Chen, R.; Zheng, Z.; Li, N.; Li, J.; Feng, F. In-situ investigation of phase transformation behaviors of 300M steel in continuous cooling process. *Mater. Charact.* **2018**, *144*, 400–410. [CrossRef]
12. Chen, R.; Zheng, Z.; Li, J.; Li, N.; Feng, F. In Situ Investigation of Grain Evolution of 300M Steel in Isothermal Holding Process. *Materials* **2018**, *11*, 1862. [CrossRef]

13. Chen, R.; Hong, C.; Li, J.; Zheng, Z.; Li, P.C. Austenite grain growth and grain size distribution in isothermal heat-treatment of 300M steel. *Procedia Eng.* **2017**, *207*, 663–668. [CrossRef]
14. Liu, Y.G.; Li, M.Q.; Luo, J. The modelling of dynamic recrystallization in the isothermal compression of 300M steel. *Mater. Sci. Eng. A* **2013**, *574*, 1–8. [CrossRef]
15. Zeng, R.; Huang, L.; Li, J.; Li, H.; Zhu, H.; Zhang, X. Quantification of multiple softening processes occurring during multi-stage thermoforming of high-strength steel. *Int. J. Plast.* **2019**, *120*, 64–87. [CrossRef]
16. Semiatin, S.L. Forming and Forging. In *ASM Handbook*; ASM International: Almere, The Netherlands, 2005; Volume 14, pp. 584–585.
17. Xi, T.; Yang, C.; Shahzad, M.; Yang, K. Study of the processing map and hot deformation behavior of a Cu-bearing 317LN austenitic stainless steel. *Mater. Des.* **2015**, *87*, 303–312. [CrossRef]
18. Chen, R. A Specimen for the in situ Compression Test on the High Temperature Confocal Laser Scanning Microscope. Chinese Patent 201910327549.4, 23 April 2019.
19. Babu, K.A.; Mandal, S.; Athreya, C.N.; Shakthipriya, B.; Sarma, V.S. Hot deformation characteristics and processing map of a phosphorous modified super austenitic stainless steel. *Mater. Des.* **2017**, *115*, 262–275. [CrossRef]
20. Mohamadizadeh, A.; Zarei-Hanzaki, A.; Abedi, H.R.; Mehtonen, S.; Porter, D. Hot deformation characterization of duplex low-density steel through 3D processing map development. *Mater. Charact.* **2015**, *107*, 293–301. [CrossRef]
21. Chen, L.; Ma, X.; Liu, X.; Wang, L. Processing map for hot working characteristics of a wrought 2205 duplex stainless steel. *Mater. Des.* **2011**, *32*, 1292–1297. [CrossRef]
22. Kitahara, H.; Ueji, R.; Tsuji, N.; Minamino, Y. Crystallographic features of lath martensite in low-carbon steel. *Acta Mater.* **2006**, *54*, 1279–1288. [CrossRef]
23. Nolze, G. Irrational orientation relationship derived from rational orientation relationships using EBSD data. *Cryst. Res. Technol.* **2010**, *43*, 61–73. [CrossRef]
24. Kitahara, H.; Ueji, R.; Ueda, M.; Tsuji, N.; Minamino, Y. Crystallographic analysis of plate martensite in Fe–28.5 at.% Ni by FE-SEM/EBSD. *Mater. Charact.* **2005**, *54*, 378–386. [CrossRef]
25. Chen, G.; Fu, G.; Cheng, C.; Yan, W.; Zou, Z.; Lin, S. Effects of strain rate on dynamic recrystallization microstructure of 3003 aluminum alloy in process of hot deformation. *Trans. Mater. Heat. Treat.* **2012**, *33*, 26–31.
26. Chen, R.; Zheng, Z.; Li, J.; Fei, F. Constitutive Modeling and Hot Workability Analysis by Microstructure Examination of GH4169 Alloy. *Crystals* **2018**, *8*, 282. [CrossRef]

metals

MDPI

Article

Constitutive Equation of GH4169 Superalloy and Microstructure Evolution Simulation of Double-Open Multidirectional Forging

Yongbo Jin [1,2], Hao Xue [1], Zheyi Yang [1], Lili Zhang [3], Chunxiang Zhang [2], Sirui Wang [1] and Junting Luo [1,2,*]

1 Education Ministry Key Laboratory of Advanced Forging & Stamping Technology and Science, Yanshan University, Qinhuangdao 066004, China; ysu18233350795@163.com (Y.J.); 15667079639@163.com (H.X.); 18840871396@163.com (Z.Y.); xichenxu123@163.com (S.W.)

2 State Key Laboratory of Metastable Materials Science and Technology, Yanshan University, Qinhuangdao 066004, China; zhangchunxiang@ysu.edu.cn

3 School of Art and Design, Yanshan University, Qinhuangdao 066004, China; 18712761815@163.com

* Correspondence: luojunting@ysu.edu.cn; Tel.: +86-335-805-2253

Received: 4 September 2019; Accepted: 24 October 2019; Published: 25 October 2019

Abstract: This paper presented a double-open multidirectional forging with relatively few deformation passes and a uniform deformation. The constitutive equation and dynamic recrystallization model of the GH4169 superalloy were identified based on a thermal compression test and imported into Deform simulation software. The microstructure evolution law of GH4169 superalloy undergoing double-open multidirectional forging was simulated. The evolution of the recrystallization volume fraction and recrystallized grain size of the GH4169 superalloy during double-open multidirectional forging was obtained. Both higher temperatures and more passes were found to produce more complete recrystallization and smaller recrystallization grain size. At the maximum temperature studied, 1000 °C, with nine passes, the recrystallization volume fraction exceeded 95%, and the recrystallized grain size reached 3–5.5 μm.

Keywords: constitutive equation; GH4169 superalloy; microstructure evolution simulation; multidirectional forging

---

1. Introduction

GH4169 alloy is a nickel-based superalloy that is extensively used in a steam turbine, aerospace, chemical, and nuclear industries [1–3]. However, the original blanks produced during casting are unsatisfactory due to increased performance requirements, and thus, must be improved to meet the performance standards [4–6]. Multidirectional forging is a plastic processing method for obtaining a fine grain structure by continuously changing the direction of axial external loads, thereby compressing the forgings in different directions [7]. Multidirectional forging is a plastic processing method to obtain fine grain structure by continuously changing the direction of axes of external loads and compressing forgings in different directions. The multi-directional forging process can improve the microstructure of the material and obtain fine-grained microstructure materials with excellent mechanical properties and uniform properties. This process is particularly suitable for the regulation of the structure and properties of GH4169 alloy materials.

In recent years, research on the multidirectional forging process in the process of thermal deformation has gradually increased at home and abroad. Mikhail et al. [8] mainly studied the influence of isothermal multi-directional forging on the microstructure evolution of conventional Al-Mg-based alloys in the strain range of 1.5–6.0 and the temperature range of 200–500 °C. Xia et al. [9] conducted a multi-directional forging of Mg-Gd-Y-Nd-Zr alloy at low temperatures to study grain refinement, microstructure, and uniformity of the mechanical properties. Aoba et al. [10] systematically studied the microstructure evolution and mechanical properties of 6000 series aluminum alloys subjected to multi-directional forging and artificial aging treatment. Lin et al. [11] derived a constitutive and microstructure evolution model for GH4169 superalloy based on equivalent dislocation density. However, relevant studies on the simulation of multi-directional forging microstructures are still rare.

In this study, a double-open multidirectional forging with relatively few deformation passes and uniform deformation was generated. The constitutive equation and dynamic recrystallization model were formulated based on a thermal compression experiment on the GH4169 superalloy. The model was then imported into the DEFORM simulation software. The design process of double-open multidirectional forging was explored, and specifically, the predictions of the recrystallization volume fraction and grain size as functions of temperature during the microstructure evolution simulation of GH4169 superalloy were discussed.

2. Experimental Materials and Methods

The material used in the experiment was a GH4169 alloy cylinder (φ 6 mm × 9 mm). The original microstructure of the sample was approximately uniform equiaxed, and the average grain size was approximately 25 μm, as shown in Figure 1. Heat treatment is necessary for the dynamic-recrystallization hot compression test of the samples. The equipment used in the experiment was the Gleeble-3800 Thermal Simulation Tester, and the specific process parameters are shown in Table 1. Figure 2 shows a flowchart of the GH4169 superalloy hot compression test. For the Gleeble-3800 thermal compression test, the alloy samples were initially heated at a rate of 5 °C/s to the required temperature (900 °C, 950 °C, 1000 °C, 1050 °C, or 1120 °C) for the experiment. Once the desired temperature was maintained for 2 min, the sample was subjected to a hot compression test at different strain rates (0.001, 0.01, 0.1, and 1 s^{-1}). Each sample was immediately water-cooled at the end of the experiment [12–14].

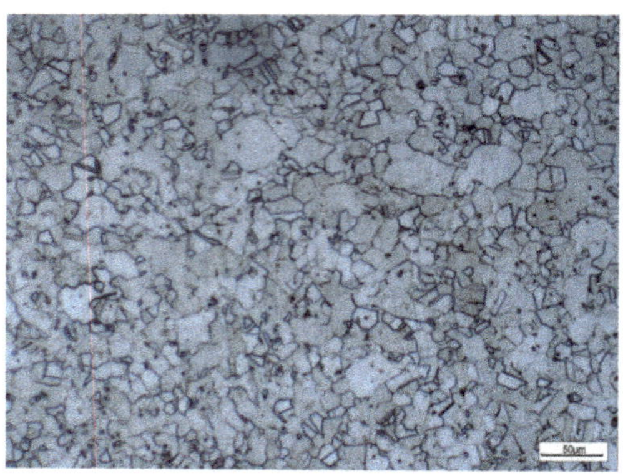

Figure 1. The original microstructure of the GH4169 forge piece.

Table 1. Deformation parameters for the GH4169 superalloy hot compression experiment.

Temperature/°C	900	950	1000	1050	1120
Strain rate/s^{-1}	0.001, 0.01, 0.1, 1				
Heating rate/°C·s^{-1}	5				
Compression ratio	0.6				

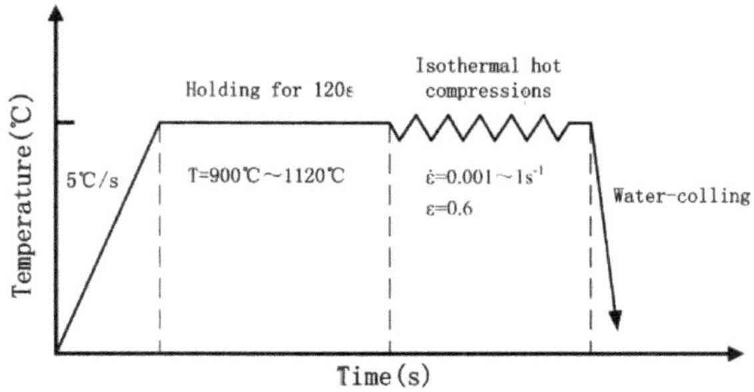

Figure 2. Flowchart of the GH4169 superalloy hot compression experiment.

3. Results and Discussion

3.1. True Stress-Strain Curve at Different Strain Rates

Figure 3 shows the true stress-strain curve obtained for the GH4169 superalloy at different strain rates and temperatures. The stress gradually decreased with an increase in temperature at a given strain and strain rate. Both the peak stress and strain point corresponding to the peak stress gradually decreased with increasing temperature. The material had a similar response to thermal deformation at different strain rates; that is, the stress value increased with the deformation amount. Meanwhile, the stress value decreased with an increase in deformation when the stress value reached its maximum (peak stress). Furthermore, the stress of the material remained constant with the increase in strain when the stress value reached steady-state stress.

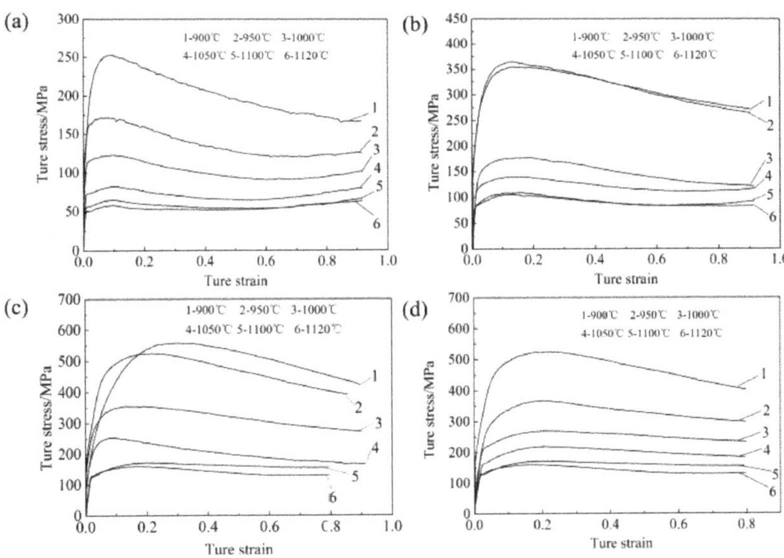

Figure 3. Thermal compression strain curve of GH4169 alloy under different strain rates and temperatures. (**a**) $\dot{\varepsilon} = 0.001\text{s}^{-1}$; (**b**) $\dot{\varepsilon} = 0.01\text{s}^{-1}$; (**c**) $\dot{\varepsilon} = 0.1\text{s}^{-1}$; (**d**) $\dot{\varepsilon} = 1\text{s}^{-1}$.

The sharp increase of stress in the initial stage of thermal compression deformation was due to work hardening. Although recovery and recrystallization might occur, work hardening played a leading role. Because of the slippage of grains in the process of plastic deformation and the tangling of dislocations, the grains were elongated or broken and fibrozed, which resulted in the formation of residual stress in the metal. As the deformation continued to increase, the stress was gradually reduced due to work softening. The subsequent stress reduction during deformation was mainly due to recovery and recrystallization, and work softening caused by recovery and recrystallization played a leading role in this. The stress gradually approached a stable value and remained constant with an increase in deformation. At this point, the work hardening and work softening of the material were balanced and entered a steady deformation stage [15,16].

The stress and peak stress gradually decreased with an increase in the strain rate under certain temperature conditions. The strain point corresponding to the peak stress increased with the temperature and amount of deformation. The deformation showed the same response at different temperatures; that is, the stress began to rapidly increase due to work hardening and reached a peak as the amount of deformation increased. Then, the stress gradually decreased and stabilized due to the influence of work softening [17,18].

3.2. Construction of Constitutive Equations

The flow stress of the metal during hot deformation is related to the thermodynamic parameter Z and deformation rate $\dot{\varepsilon}$ of the material. The relation is expressed as follows [19,20]:

$$Z = \dot{\varepsilon}exp(\frac{Q}{RT}), \quad (1)$$

where Q is the thermal deformation activation energy of the material, R is the gas constant, and T is the absolute temperature.

Materials have different stress function forms under different conditions, as follows:

$$F(\sigma) = A_1\sigma^n(\alpha\sigma < 0.8) \quad (2)$$

$$F(\sigma) = A_2exp(\beta\sigma)(\alpha\sigma > 1.2) \quad (3)$$

$$F(\sigma) = A_2exp(\beta\sigma)(\alpha\sigma > 1.2) \quad (4)$$

where A_i (I = 1, 2, 3); α, β, and n are the material parameters.

The sinusoidal equation proposed by Sellars et al. is generally used to describe the relationship between various material parameters during plastic deformation and is expressed as follows:

$$\dot{\varepsilon} = A[sinh(\alpha\sigma)]^n exp\left[-\frac{Q}{RT}\right] \quad (5)$$

where α, n, A, and β are the material parameters. Q is the thermal deformation activation energy of the material. R is the gas constant, which is a fixed value (R = 8.31). $\dot{\varepsilon}$ is the deformation rate. Σ is the flow stress, and T denotes absolute temperature.

Substituting Equations (2) and (3) into Equation (6) and deriving on two sides of the equation, Equation (7) could be obtained.

$$ln\dot{\varepsilon} = lnA_1 + nln\sigma - \frac{Q}{RT} \quad (6)$$

$$ln\dot{\varepsilon} = lnA_2 + \beta\sigma - \frac{Q}{RT} \quad (7)$$

For convenience, the logarithmic form of the equation is often employed. Then, the corresponding graphs are made with the coordinates of $ln\dot{\varepsilon}$ and $ln\sigma$ and $ln\ \dot{\varepsilon}$ and σ, respectively. As shown in Figure 4, the peak stress corresponding to different strain rates at different temperatures was calculated. The data are also shown in Table 2.

The diagram in Figure 4 could be fitted, where the average slope of each $ln\dot{\varepsilon}$-$ln\sigma_p$ line could be calculated based on Equation (6) and Figure 4a, and the $ln\dot{\varepsilon}$-$ln\sigma_p$ lines could be calculated based on Equation (7) and Figure 4b. The resulting average slopes were n = 5.954252 and β = 0.031217. Given α = β/n, α = 0.0052428 could also be obtained.

Table 2. Strain rate ($\dot{\varepsilon}$), peak stress (σ_p), and corresponding logarithmic value of the GH4169 superalloy at various temperatures.

Temperature T/°C	Strain Rate ($\dot{\varepsilon}$)/s^{-1}	Peak Stress (σ_p)/Mpa	$ln\dot{\varepsilon}$	$ln\sigma_p$	$ln[sinh(\alpha\sigma_p)]$
900	0.001	253.1	−6.908	5.533785	0.5608306
	0.01	364.32	−4.605	5.898033	1.1947405
	0.1	523.39	−2.303	6.260327	2.0467375
	1	556.85	0	6.322296	2.2233897
950	0.001	212.46	−6.908	5.358754	0.3067085
	0.01	354.28	−4.605	5.870088	1.139611
	0.1	450.13	−2.303	6.109536	1.6578382
	1	528.47	0	6.269986	2.0735863
1000	0.001	171.81	−6.908	5.146389	0.0272396
	0.01	347.85	−4.605	5.851771	1.1041569
	0.1	367.16	−2.303	5.905798	1.2102875
	1	500.3	0	6.215208	1.9245429
1050	0.001	82.3	−6.908	4.410371	−0.809689
	0.01	139.68	−4.605	4.939354	−0.22371
	0.1	218.02	−2.303	5.384587	0.3426757
	1	303.52	0	5.715448	0.8557846
1100	0.001	74.45	−6.908	4.310128	−0.915507
	0.01	129.43	−4.605	4.86314	−0.31216
	0.1	178.39	−2.303	5.183972	0.0748297
	1	233.89	0	5.454851	0.443079
1120	0.001	66.847	−6.908	4.202406	−1.028105
	0.01	105.18	−4.605	4.655673	−0.54505
	0.1	158.97	−2.303	5.068716	−0.06898
	1	215.31	0	5.372079	0.3252004

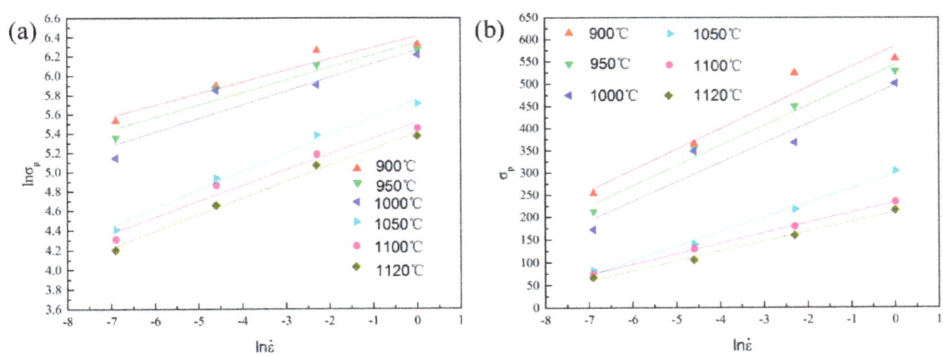

Figure 4. Relation among $ln\sigma_p$, σ_p, and $ln\dot{\varepsilon}$ curves at different temperatures. (**a**) $ln\sigma_p$ and $ln\dot{\varepsilon}$ curves at different temperatures; (**b**) σ_p and $ln\dot{\varepsilon}$ curves at different temperatures.

Deriving on both sides of Formula (6):

$$ln\dot{\varepsilon} = lnA - nln[sinh(\alpha\sigma)] - \frac{Q}{RT} \quad (8)$$

Performing partial derivation of $1/T$ at a certain strain rate for Equation (8) yields:

$$Q = R\left[\frac{\partial ln\dot{\varepsilon}}{\partial ln[sinh(\alpha\sigma)]}\right]_T\left[\frac{\partial ln[sinh(\alpha\sigma)]}{\partial(1/T)}\right] \quad (9)$$

The activation energy of the material is constant when the strain rate is constant. The n values fitted in Figure 4 were substituted into Equation (9) to obtain $ln\dot{\varepsilon}$ and $ln[sinh(\alpha\sigma)]$ at different deformation temperatures and $ln[sinh(\alpha\sigma)]$ and 1000/T at different strain rates, where 1000 times 1/T was used for convenient calculation, as shown in Figures 5 and 6.

The average slope k = 0.231755 of each straight line could be calculated based on the straight line fitted in Figure 5, and the average slope of each straight line t = 13.90354 could be calculated based on the straight line fitted in Figure 6. Based on Equation (9), the thermal deformation energy of the material could be calculated as follows:

$$Q = \frac{R*T}{k} = 498.54\frac{kJ}{mol} = 498540\ J/mol \tag{10}$$

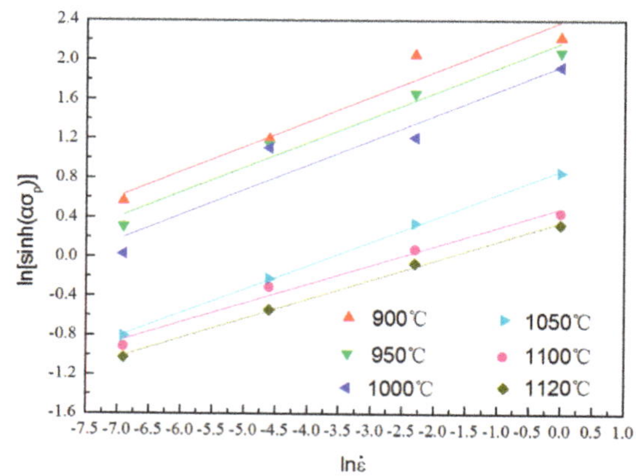

Figure 5. The relation between $ln[sinh(\alpha\sigma_p)]$ and $ln\dot{\varepsilon}$ at various deformation temperatures.

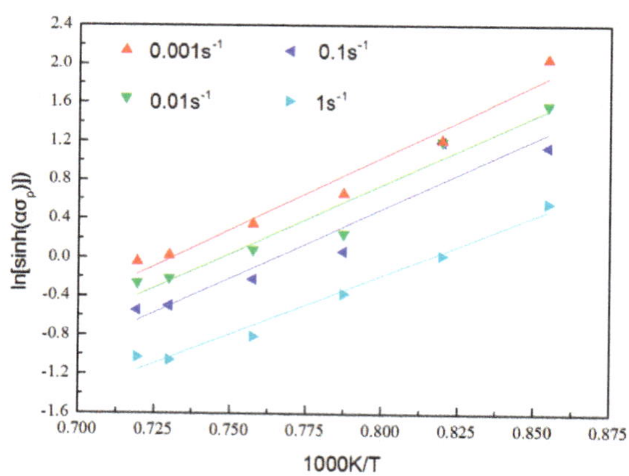

Figure 6. The relation between $ln[\sinh(\alpha\sigma_p)]$ and 1000 K/T at various strain rates.

From Equations (1) and (6), Equation (11) was obtained:

$$Z = \dot{\varepsilon}exp\left(\frac{Q}{RT}\right) = A[sinh(\alpha\sigma)]^n \tag{11}$$

Performing derivation on both sides of Equation (11):

$$lnZ = ln\dot{\varepsilon} + \frac{Q}{RT} \tag{12}$$

$$lnZ = lnA + nln[sinh(\alpha\sigma)] \tag{13}$$

The corresponding value of lnZ could be calculated by comparing the peak stress values for different temperatures and strain rates, and the corresponding lnZ and $ln[sinh(\alpha\sigma_p)]$ maps were fitted using Equation (13), as shown in Figure 7. The slope of the line was n = 4.56568, and the intercept was lnA = 42.21596 ($A = 2.1585 \times 10^{18}$). Thus, $Z = 2.1585 \times 10^{18} \times [sinh(0.0052428\sigma)]^{4.56568}$ was obtained.

Substituting all e values obtained into Equation (6) yielded the constitutive equation of the alloy, as shown in Equation (14).

$$\dot{\varepsilon} = 2.1585 \times 10^{18} \times [sinh(0.0052428\sigma)]^{4.56568} \times exp(-\frac{498540}{RT}) \quad (14)$$

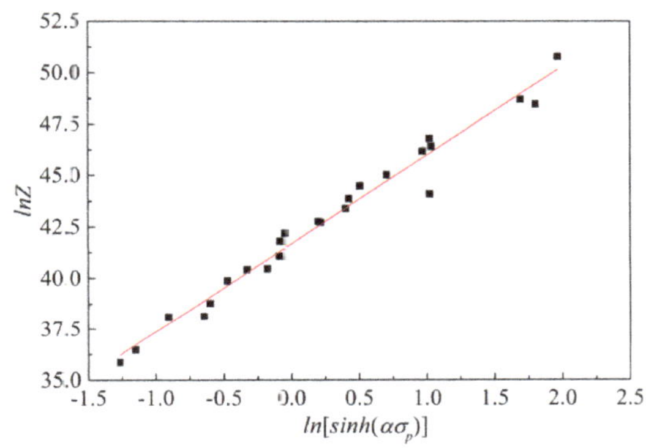

Figure 7. Relation between lnZ and $ln[sinh(\alpha\sigma_p)]$ at various strain rates and deformation temperatures.

3.3. Dynamic Recrystallization Model

3.3.1. Proposed Model

Existing studies generally use the Avrami equation to describe the recrystallization degree quantitatively [21,22]:

$$X = 1 - exp[-k(\frac{\varepsilon - \varepsilon_c}{\varepsilon_{0.5}})^n] \quad (15)$$

where X is the dynamically recrystallized volume fraction of the material; k and n are the material parameters; ε is a dependent variable; ε_c is the critical strain; $\varepsilon_{0.5}$ is the recrystallized amount of the material when 50% strain is reached.

Peak strain model:

$$\varepsilon_p = AZ^m \quad (16)$$

where A and m are the material parameters; Z is the temperature compensation factor.

Critical strain model:

$$\varepsilon_c = k\varepsilon_p \quad (17)$$

where k ranges from 0.6 to 0.85, and a value of 0.8 was used in this study.

A quantitative description of the recrystallization quality is usually given by:

$$D_{2-drex} = A_1 Z^{A_2} \quad (18)$$

where A_1 and A_2 are material-dependent constants.

3.3.2. Model Establishment

The peak strain value of the material during hot deformation could be obtained through experiments, and the corresponding recrystallization volume fraction and lnZ value were calculated, as shown in Table 3.

Simplifying Equation (16) yielded:

$$ln\varepsilon_p = lnA + mlnZ \tag{19}$$

As shown in Equation (18), the corresponding $ln\varepsilon_p$ and lnZ maps could be constructed and fitted (Figure 8). The intercept of the line was $lnA = -4.46775$, that is, $A = 1.15 \times 10^{-2}$, and the slope of the line was $m = 0.067$.

Substituting the obtained A and m values into Equation (16) yielded Equation (20).

$$\varepsilon_p = 1.15 \times 10^{-2} Z^{0.067} \tag{20}$$

Then, by Equation (17):

$$\varepsilon_c = 9.2 \times 10^{-3} Z^{0.067} \tag{21}$$

The strain value corresponding to material recrystallization of 50% could be obtained by performing Newton interpolation on the parameter values obtained from the experiment, as shown in Equation (22).

$$\varepsilon_{0.5} = 0.29Z^{0.016} \tag{22}$$

Simplifying Equation (15) yielded:

$$ln\left[ln\left(\frac{1}{1-X}\right)\right] = lnk + nln(\frac{\varepsilon - \varepsilon_c}{\varepsilon_{0.5}}) \tag{23}$$

The corresponding $ln\{ln[1/(1 - X)]\}$ and $ln[(\varepsilon - \varepsilon_c)/\varepsilon_{0.5}]$ maps could be constructed from Equation (23), and a linear fit could be performed, as shown in Figure 9. By fitting the data in Figure 9, one found $k = 0.812$ and $n = 0.92$.

Substituting the values of k and n into Equation (15) yielded:

$$X = 1 - exp[-0.812(\frac{\varepsilon - \varepsilon_c}{\varepsilon_{0.5}})^{0.92}] \tag{24}$$

Table 3. Strain rate $\dot{\varepsilon}$, peak strain ε_p, dynamic recrystallization volume fraction X, and lnZ at different temperatures.

Temperature/°C	Srain Rate $\dot{\varepsilon}$/s^{-1}	Peak Strain ε_p/MPa	lnZ	X
950	0.001	0.221	42.26602	0.724
	0.01	0.156	40.33004	0.658
	0.1	0.105	38.54073	0.543
	1	0.113	36.88203	0.287
1000	0.001	0.1	36.25196	0.927
	0.01	0.228	44.5686	0.843
	0.1	0.2	42.63263	0.706
	1	0.168	40.84331	0.504
1050	0.001	0.154	39.18461	0.895
	0.01	0.144	38.55454	0.919
	0.1	0.232	46.87119	0.817
	1	0.221	44.93521	0.665
1100	0.001	0.206	43.1459	0.905
	0.01	0.188	41.4872	0.928
	0.1	0.18	40.85713	0.886
	1	0.237	49.17377	0.780
1120	0.001	0.233	47.2378	0.887
	0.01	0.225	45.44848	0.928
	0.1	0.212	43.78978	0.904
	1	0.206	43.15971	0.815

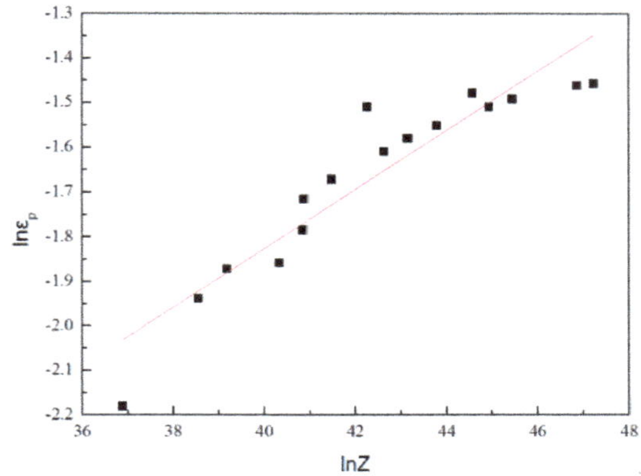

Figure 8. The relation between lnZ and $ln\varepsilon_p$ at various deformation temperatures.

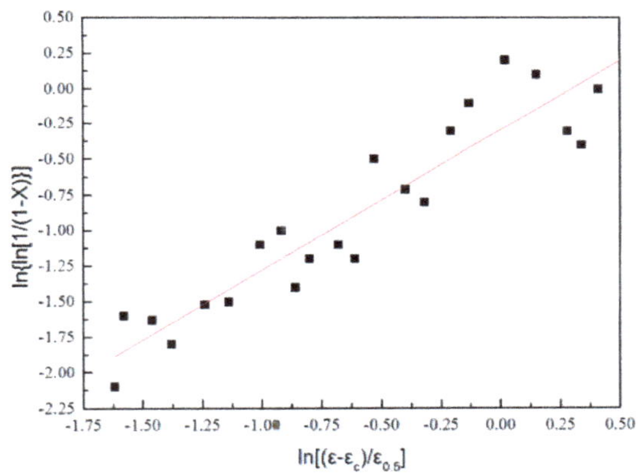

Figure 9. Relation between $ln\{ln[1/(1-X)]\}$ and $ln[(\varepsilon-\varepsilon_c)/\varepsilon_{0.5}]$ at various deformation temperatures.

The grain size of the microstructure calculated according to the GH4169 high-temperature dynamic recrystallization metallographic structure diagram is shown in Table 4. According to the data in the table, the logarithmic fitting of $D_{2\text{-}drex}$ and Z could be used to find the corresponding model parameters. The fitted image is shown in Figure 10. The obtained dynamic recrystallization crystal mass equation is shown in Equation (25).

$$D_{2-drex} = 3\times10^{6}Z^{-0.3016} \tag{25}$$

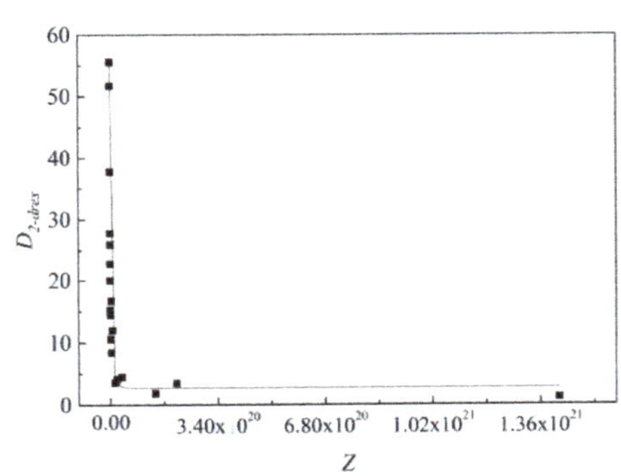

Figure 10. $D_{2\text{-}drex}$ and Z fitted image.

Table 4. Strain rate $\dot{\varepsilon}$, dynamic recrystallization grain size, and Z of the GH4169 superalloy at various temperatures.

Temperature/°C	Strain Rate/s^{-1}	Z	$D_{2\text{-}drex}$/μm
950	0.001	1.42×10^{18}	10.556
	0.01	1.42×10^{19}	3.578
	0.1	1.42×10^{20}	1.818
	1	1.42×10^{21}	1.111
1000	0.001	2.09×10^{17}	22.78
	0.01	2.09×10^{18}	10.603
	0.1	2.09×10^{19}	4.167
	1	2.09×10^{20}	3.442
1050	0.001	3.57×10^{16}	37.78
	0.01	3.57×10^{17}	15.2
	0.1	3.57×10^{18}	8.425
	1	3.57×10^{19}	4.487
1100	0.001	6.93×10^{15}	51.67
	0.01	6.93×10^{16}	25.92
	0.1	6.93×10^{17}	14.444
	1	6.93×10^{18}	12
1120	0.001	3.72×10^{15}	55.56
	0.01	3.72×10^{16}	27.78
	0.1	3.72×10^{17}	20
	1	3.72×10^{18}	16.667

3.4. Double-Open Multidirectional Forging Simulation

3.4.1. Process and Finite Element Model

The finite element model used in the simulation is shown in Figure 11. During forging, the initial billet was compressed in the height direction, stretched in the longitudinal direction, and did not deform in the width direction, owing to the restraining effect of the mold. When the amount of deformation reached a certain level, the closed multidirectional forging occurred due to the restraining action of the lower die, the length of the forging was no longer increased in the longitudinal direction, and the groove of the lower die was gradually filled. Meanwhile, the single-open multidirectional forging side was full, the other side maintained a free surface, and the sides of the double-open multidirectional forging maintained a free surface. The forging was rotated by 90° every time it was swept, and the last forging was repeated. The simulation utilized a relative net partitioning method and a tetrahedral mesh with 20,000 cells. Table 5 shows the parameter settings used for the double-open multidirectional forging simulation.

Table 5. Parameter list.

Process Parameters	Symbol	Unit	Value
Forging size	-	mm	40 × 40 × 50
Mold material	-	-	H13
Forging temperature	T	°C	800–1000
Coefficient of friction	f	-	0.3
Thermal conductivity	λ	W/(m·K)	20–40
Specific heat	C	N/(mm^2·K)	3–5
Initial grain size	d	μm	45
Upper die size	-	mm	50 × 40 × 10
Lower die size	-	mm	70 × 60 × 60
Reduction rate	-	%	20

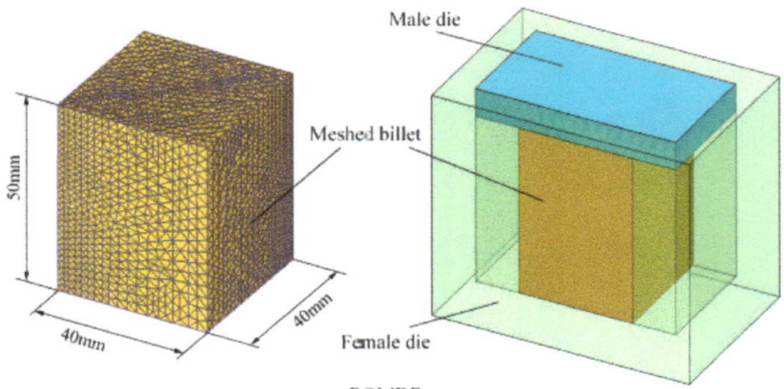

Figure 11. Finite element model.

3.4.2. Numerical Simulation Results and Analysis

Figure 12 shows the dynamic recrystallization volume fraction cloud diagram of forgings from three to nine passes at 800 °C. The dynamic recrystallization volume fraction of the forging was symmetrically distributed and could be divided into three deformation zones, namely difficult, easy, and free. The equivalent strain was minimized on the outer surface of the forging due to the restraint of the mold. Therefore, the recrystallization degree in this area was relatively low, the area with the largest recrystallization was approximately 30%, and some areas were not recrystallized. In the interior of the forging, the equivalent strain was large, and the recrystallization degree was relatively high because it was not affected by the mold friction. The recrystallization degree near the inner forging was high, and the center portion reached approximately 82%.

Figure 13 shows the recrystallized grain size graph of forging from three to nine passes at 800 °C. The cloud diagram shows that the recrystallized grain size of the forging was symmetrically distributed. The closer it was to the forging center, the smaller was the grain size. The equivalent strain of the forging was also small on the outer surface of the forging due to the limitation of the die on the forging, thereby affecting the recrystallization. Thus, the average grain size of the forging in this area was large, and no recrystallization occurred. The equivalent strain in the interior of the forging was large because it was not affected by the mold friction. Thus, the average grain size was small, with a minimum value of approximately 33.5 μm.

The recrystallization degree of the forgings was remarkably improved when it reached six passes. The closer it was to the internal, the higher was the recrystallization degree, which resulted in the center portion to approximately reach 95.5%. Therefore, the recrystallization degree was high, and the average grain size was small. The minimum value was approximately 12.2 μm. When forging nine passes, the recrystallization degree of the forgings reached 98.5%, and the minimum average grain size was approximately 5.1 μm.

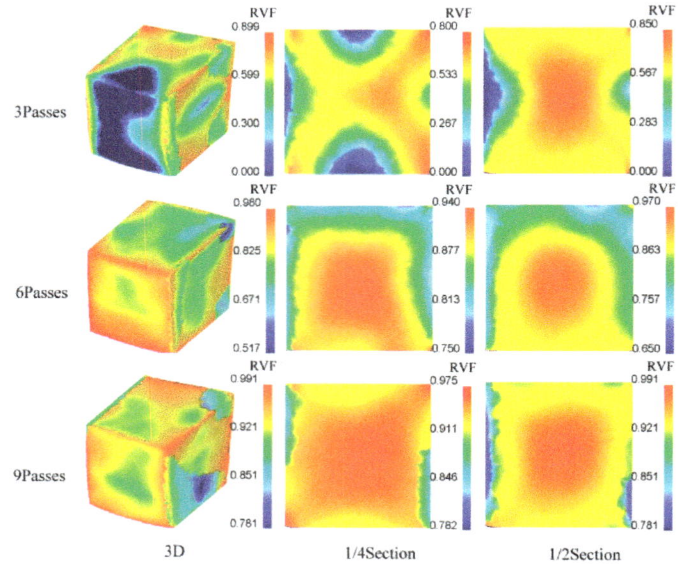

Figure 12. Dynamic recrystallization volume fraction cloud diagram for three to nine passes forging at 800 °C.

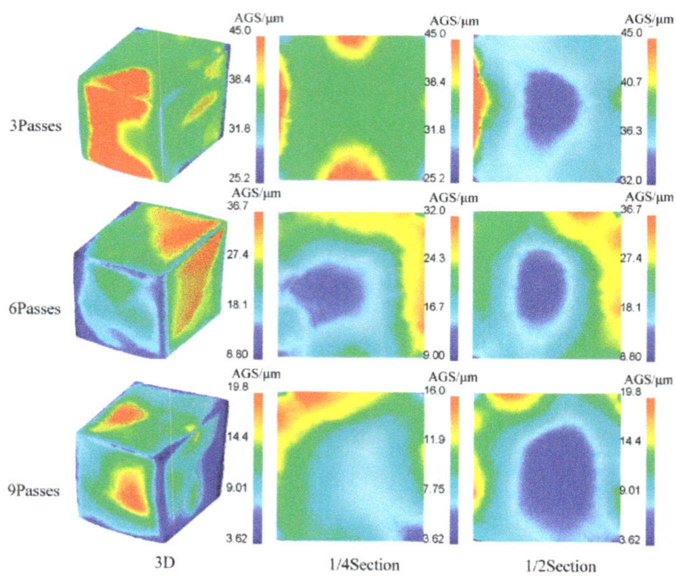

Figure 13. Forging of three to nine passes for recrystallized grain size cloud map at 800 °C.

Figure 14 shows the dynamic recrystallization volume fraction cloud diagram of forging from three to nine passes at 900 °C. From the figure, the dynamic recrystallization volume fraction of the forging was symmetrically distributed and could be divided into three deformation zones: difficult, easy, and free deformation. The equivalent strain was minimized on the outer surface due to the mold constraint. Therefore, the recrystallization degree in this area was relatively low, the area with the largest recrystallization was approximately 50%, and some areas were not recrystallized. In the interior of the forging, because it was not affected by the mold friction, the equivalent strain was large;

the recrystallization degree was relatively high, and the closer it was to the internal, the higher was the recrystallization degree, which resulted in the center portion achieving a value of approximately 87%.

Figure 15 shows the recrystallized grain size cloud diagram of forging from three to nine passes at 900 °C. From the figure, the recrystallized grain size of the forging was symmetrically distributed, and the closer it was to the inside of the forging, the smaller was the grain size, and the closer the outer grain size was, the larger was the grain size. The equivalent strain was small on the outer surface of the forging due to the limitation of the die on the forging, which affected its recrystallization. Thus, the average grain size of the forging in this area was large, and no recrystallization occurred. In the interior of the forging, because it was not affected by the mold friction, the equivalent strain was large. Thus, the recrystallization degree in this region was high, the average grain size was small, and the minimum value was approximately 26 μm.

The center portion reached approximately 97.5% under the forging of six passes. Therefore, the recrystallization degree was high in this region, the average grain size was small, and the minimum value was approximately 10 μm. When forging nine passes, the recrystallization degree of the forgings reached approximately 99.7%, and the average grain size minimum was approximately 4 μm.

Figure 16 shows the dynamic recrystallization volume fraction cloud diagram of forgings from three to nine passes at 1000 °C. From the figure, the dynamic recrystallization volume fraction of the forging was symmetrically distributed and could be divided into three deformation zones: difficult, easy, and free deformation zones. The equivalent strain was minimized on the outer surface of the forging due to the constraint of the mold. Thus, the recrystallization degree in this area was relatively low, and the area with the largest recrystallization was approximately 70%. In the interior of the forging, because it was not affected by the mold friction, the equivalent strain was large, and the recrystallization degree was relatively high. The closer it was to the internal, the higher was the recrystallization degree, which resulted in the highest value of approximately 74% at the center; moreover, the degree of internal recrystallization did not change considerably with the outside. No obvious increasing trend was observed.

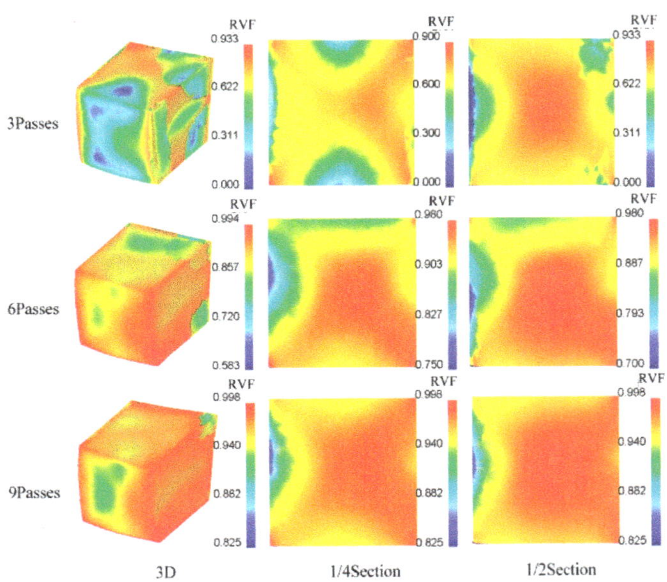

Figure 14. Forging from three to nine passes for dynamic recrystallization volume fraction cloud image at 900 °C.

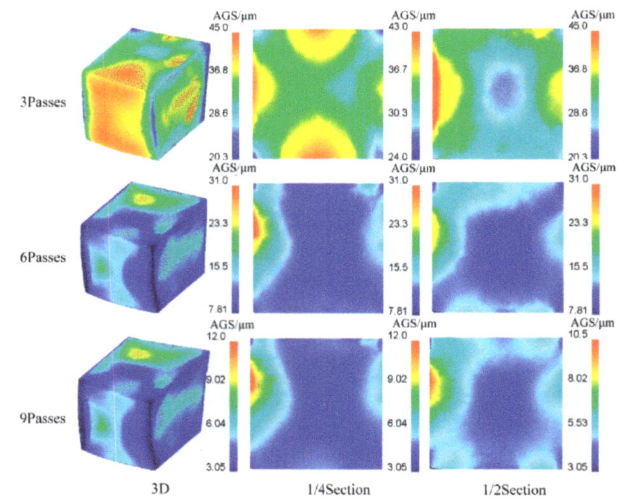

Figure 15. Forging from three to nine passes for recrystallized grain size cloud image at 900 °C.

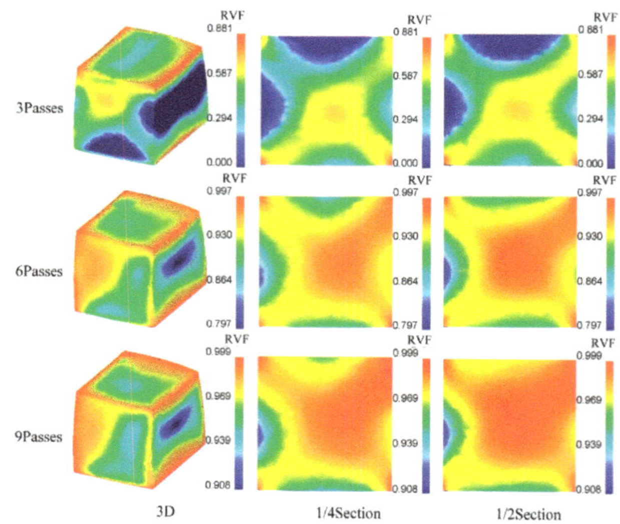

Figure 16. Dynamic recrystallization volume fraction cloud image of forging from three to nine passes at 1000 °C.

Figure 17 shows the recrystallized grain size graph of forging from three to nine passes at 1000 °C. From the figure, the recrystallized grain size of the forging was symmetrically distributed; moreover, the closer it was to the inside of the forging, the smaller was the grain size, and the closer it was to the outer grain size, the larger was the grain size. The equivalent strain was small on the outer surface of the forging due to the restraining effect of the die on the forging, which affected the recrystallization of the forging. The average grain size of the forging in this area was relatively large. In the interior of the forging, because it was not affected by the mold friction, the equivalent strain was large. Thus, the recrystallization degree in this area was relatively high, the average grain size was small, the minimum value was approximately 24 μm, and the internal and external grain sizes of the forging did not change significantly. This situation occurred because the deformation temperature of the forging was relatively high, and the strain had no considerable effect on recrystallization.

The center portion approximately reached 99% with the forging of six passes. Therefore, the recrystallization degree was high in this region, the average grain size was small, and the minimum value was approximately 9.1 μm. When forging nine passes, all areas of recrystallization volume fraction exceeded 95%, the grain size was 3–5.5 μm, the recrystallization degree of forgings reached 99.7%, and the average grain size minimum was approximately 3 μm.

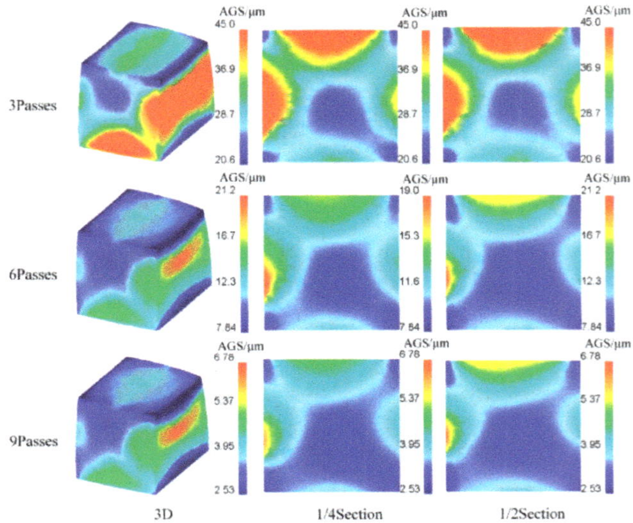

Figure 17. Recrystallization grain size cloud image of forging from three to nine passes at 1000 °C.

4. Conclusions

(1) Based on the thermal compression experiment, the constitutive model of the material and the mathematical model of recrystallization evolution were derived by analyzing the stress-strain curve of GH4169 and applied to the secondary development of software. A simulation of the microstructure evolution law in the double-open multi-directional forging process was conducted.

(2) The evolution rule of the recrystallization volume fraction and the recrystallized grain size of the GH4169 superalloy double-open multidirectional forging was obtained. Increasing the forging temperature increased the recrystallization volume fraction and reduced the recrystallization grain size. Performing more passes had the same effect and led to more complete recrystallization and a smaller recrystallized grain size. At 1000 °C and nine passes, the recrystallization volume fraction exceeded 95%, and the recrystallized grain size reached the minimum size of 3–5.5 μm.

Author Contributions: J.L. and Y.J. conceived and designed the experiments; H.X., S.W., and Z.Y. performed the experiments; L.Z. and C.Z. analyzed the data; J.L. and Y.J. wrote the paper.

Funding: This project is supported by Natural Science Foundation of Hebei Province, China (Grant No. E2019203005).

Conflicts of Interest: The authors declare no conflict of interest.

References

1. Zhou, L.X.; Baker, T.N. Eeffct of strain rate and temperuatre on deofrmation behvaior of IN718 during high temperuatre deformation. *Mater. Sci. Eng.* **1994**, *177*, 1–9. [CrossRef]
2. Du, J.H.; Lu, X.D.; Deng, Q.; Bi, Z.B. Progress in the Research and Manufacture of GH4169 alloy. *J. Iron Steel Res. Int.* **2015**, *22*, 657–663. [CrossRef]

3. Li, H.Y.; Kong, Y.H.; Chen, G.S.; Xie, L.X.; Zhu, S.G.; Sheng, X. Effect of different processing technologies and heat treatments on the microstructure and creep behavior of GH4169 superalloy. *Mater. Sci. Eng.* **2013**, *582*, 368–373. [CrossRef]
4. Zhou, X.; Zhang, J.; Chen, X.; Zhang, X.; Li, M. Fabrication of high-strength AZ80 alloys via multidirectional forging in air with no need of ageing treatment. *J. Alloys Compd.* **2019**, *787*, 551–559. [CrossRef]
5. Na, Y.S.; Yeom, J.T.; Park, N.K.; Lee, J.Y. Simulation of microstructures for Alloy 718 blade forging using 3D FEM simulator. *J. Mater. Process. Technol.* **2003**, *141*, 337–342. [CrossRef]
6. Salevati, M.A.; Akbaripanah, F.; Mahmudi, R. Microstructure, Texture, and Mechanical Properties of AM60 Magnesium Alloy Processed by Extrusion and Multidirectional Forging. *J. Mater. Eng. Perform.* **2019**, *28*, 3021–3030. [CrossRef]
7. Miura, H.; Maruoka, T.; Yang, X.; Jonas, J.J. Microstructure and mechanical properties of multi-directionally forged Mg–Al–Zn alloy. *Scr. Mater.* **2012**, *66*, 49–51. [CrossRef]
8. Kishchik, M.S.; Mikhaylovskaya, A.V.; Kotov, A.D.; Mosleh, A.O.; AbuShanab, W.S.; Portnoy, V.K. Effect of multidirectional forging on the grain structure and mechanical properties of the Al–Mg–Mn alloy. *Materials* **2018**, *11*, 2166. [CrossRef]
9. Xia, X.S.; Wang, C.P.; Wu, Y.; Zhang, K.; Ma, M.L.; Yuan, B.G. Grain refinement, microstructure and mechanical properties homogeneity of Mg-Gd-Y-Nd-Zr alloy during multidirectional forging. *J. Mater. Eng. Perform.* **2018**, *27*, 5689–5699. [CrossRef]
10. Aoba, T.; Kobayashi, M.; Miura, H. Microstructural evolution and enhanced mechanical properties by multi-directional forging and aging of 6000 series aluminum alloy. *Mater. Trans.* **2018**, *59*, 373–379. [CrossRef]
11. Lin, J.; Liu, Y. A set of unified constitutive equations for modellling microstructure evolution in hot deformation. *J. Mater. Process. Technol.* **2003**, *143*, 281–285. [CrossRef]
12. Kermanpur, A.; Lee, P.D.; Mclean, M.; Tin, S. Integrated modeling for the manufacture of aerospace discs: Grain structure evolution. *JOM* **2004**, *56*, 72–78. [CrossRef]
13. Tin, S.; Lee, P.D.; Kermanpur, A.; Mclean, M.; Rist, M. Integrated modeling for the manufacture of Ni-based superalloy discs from solidification to final heat treatment. *Metall. Mater. Trans. A* **2005**, *36*, 2493–2504. [CrossRef]
14. Park, N.K.; Kim, I.S.; Na, Y.S.; Yeom, J.T. Hot forging of a nickel-base superalloy. *J. Mater. Process. Technol.* **2001**, *111*, 98–102. [CrossRef]
15. Murty, S.V.S.N.; Rao, B.N. On the development of instability criteria during hotworking with reference to IN 718. *Mater. Sci. Eng.* **1998**, *254*, 76–82. [CrossRef]
16. Cao, F.R.; Zhang, J.; Ding, X.; Xue, G.Q.; Liu, S.Y.; Sun, C.F. Mechanical properties and microstructural evolution in a superlight Mg-6.4Li-3.6Zn-0.37Al-0.36Y alloy processed by multidirectional forging and rolling. *Mater. Sci. Eng.* **2019**, *760*, 377–393. [CrossRef]
17. Marty, B.; Guedou, J.; Gergaud, P.; Lebrun, J.L. Recrystallization and work-hardening prediction during forging process of inconel 718. *Superalloys* **1997**, *718*, 625–706.
18. Zhang, Z.X.; Qu, S.J.; Feng, A.H.; Hu, X.; Shen, J. Microstructural mechanisms during multidirectional isothermal forging of as-cast Ti-6Al-4V alloy with an initial lamellar microstructure. *J. Alloys Compd.* **2019**, *773*, 277–287. [CrossRef]
19. Zhang, J.M.; Gao, Z.Y.; Zhuang, J.Y.; Zhong, Z.Y. Mathematical modeling of the hot deformation behavior of superally IN718. *Metall. Mater. Trans. A* **1999**, *30*, 2701–2713. [CrossRef]
20. Thomas, A.; El-Wahabi, M.; Cabrera, J.M.; Prado, J.M. High temperature deformation of Inconel 718. *J. Mater. Process. Technol.* **2006**, *177*, 469–472. [CrossRef]
21. Hu, J.P.; Zhuang, J.Y.; Du, J.H.; Deng, Q.; Feng, D. Constitutive equation of superalloy In718 in hammer forging process. *J. Iron Steel Res. Int.* **2001**, *8*, 50–54.
22. Dandre, C.A.; Roberts, S.M.; Evans, R.W.; Reed, R.C. Microstructural evolution of Inconel* 718 during ingot breakdown: Process modelling and validation. *Mater. Sci. Technol.* **2000**, *16*, 14–25. [CrossRef]

metals

MDPI

Article

Prediction of Earing of Cross-Rolled Al Sheets from {h00} Pole Figures

Marton Benke *, Bence Schweitzer, Adrienn Hlavacs and Valeria Mertinger

Institute of Physical Metallurgy, Metalforming and Nanotechnology, University of Miskolc, 3515 Miskolc, Hungary; thebencyx@gmail.com (B.S.); femhadri@uni-miskolc.hu (A.H.); femvali@uni-miskolc.hu (V.M.)

* Correspondence: fembenke@uni-miskolc.hu

Received: 5 December 2019; Accepted: 21 January 2020; Published: 28 January 2020

Abstract: The plastic anisotropy of rolled Al sheets is the result of a crystallographic texture. It leads to the formation of uneven cup heights during deep-drawing, which is called earing. A new, simple and rapid method had been previously developed by the authors to predict earing directly from {h00} pole figures. In the present manuscript, this method is applied to cross-rolling for the first time. 5056 type aluminum sheets were unidirectionally- (conventionally) and cross-rolled from 4 to ~1 mm thickness in 6 or 12 passes. Earing was predicted from recalculated {200} pole figures obtained after X-ray diffraction texture measurements. The results were validated by deep-drawing tests. It is shown that the proposed method predicts the type (locations of ears) and magnitude of earing with satisfactory results. However, a different scaling factor must be used to calculate the magnitude of earing for cross-rolling than for unidirectional rolling even if all other parameters (including cold rolling, texture measurements, and deep-drawing) are the same. This is because the cross-rolled sheets exhibit a similar type but weaker earing compared to the unidirectionally rolled samples.

Keywords: aluminum; cross-rolling; texture; earing

1. Introduction

The term cross-rolling is used for different special types of rolling processes [1]. Here, cross-rolling means a modified version of conventional or unidirectional sheet rolling, where the sheet—and consequently, the rolling direction (RD)—is rotated by 90° about its normal direction (ND) between subsequent passes [2–5]. The outcome is that a more random orientation distribution, that is, four-fold texture, is obtained, which is usually accompanied by a lower amount of plastic anisotropy. Li et al. applied cross-rolling on magnesium sheets and concluded that large basal plane scatter was achieved, which resulted in a more random texture [2]. Wronski et al. used cross-rolling on low carbon ferritic steel and copper and found that plastic anisotropy was decreased in copper; however, it increased in the low carbon steel [3]. Huh et al. applied cross-rolling to suppress cube texture formation in 5182 aluminum after annealing and observed a close to random structure and better formability [4]. Tang et al. obtained similar results on cross-rolled AZ31 aluminum, and besides negligible earing, they found that increased ductility was also achieved by cross-rolling [5]. Based on the literature, it can be safely stated that, in general, a more random texture and decreased plastic anisotropy is expected in aluminum alloys after cross-rolling.

Besides deep-drawing tests, several methods have been established to estimate earing. Fukui and Kudo found that earing can be predicted from the Lankford value $\Delta r = (r0 + r90)/2 - r45$ [6]. Since then, methods have been developed based on mechanical response [7], crystallography [8,9], and more complex theories were established [10]. Nowadays, prediction of earing is usually performed by finite element (FE) methods [11].

Recently, a new, simple method has been developed by the authors to predict earing based on pole figure data [12]. The advantage of the method over others is that qualitative information can be obtained quickly and directly from pole figures. If quantitative data is demanded, only one deep-drawing test is required. Furthermore, it can be combined with non-destructive (sample-cutting-free) texture measurement methods [13]. Thus, it is possible to predict earing in a truly non-destructive manner, which can have high potential in cases when unique, high value objects are to be examined. The aim of the present work is to extend the developed earing prediction method to the case of cross-rolling. For this, the method was applied on unidirectionally and cross-rolled 5056 type aluminum sheets and the results were validated with deep-drawing tests.

2. Materials and Methods

Sheets made of 5056 type aluminum having an initial thickness of 4 mm have been unidirectionally- (UD) and cross-rolled (CR) with a VonRoll experimental roll stand at the University of Miskolc. Other than the mode of rolling (UD or CR), all applied parameters were the same. Two sets of samples were produced with a different number of passes. The first set of samples (three UD samples, marked as A1, A5, and A6; and three CR samples, marked as A3K, A4K, and A9K) were cold-rolled to a ~1 mm final thickness through 6 passes. Sample A4K was rotated around ND between passes clockwise, while samples A3K and A9K were rotated around ND alternately back and forth prior to passes. The reduction during the first pass was 0.6 mm, and 0.5 mm during the subsequent passes. The final thickness of the UD samples was 1.04, 1.03, and 1. 03 mm; and that of the CR samples was 1.14, 1.13, and 1.13 mm. The second set of samples, rolled with 12 passes (one UD, marked as A10; and one CR, marked as A11K) were cold-rolled to a ~0.99 and 1.10 mm final thickness, respectively. The reduction during the first pass was 0.3 mm and 0.25 during the following passes. Sample A11K was rotated around ND alternately back and forth prior to passes. Table 1 summarizes the properties of the investigated samples.

Table 1. The investigated samples and their properties. UD: unidirectionally; CR: cross-rolled.

Number of Passes	Rolling Mode	Sample	Rotation	Final Thickness, mm
6	UD	A1	-	1.04
		A5	-	1.03
		A6	-	1.03
	CR	A3K	back and forth	1.14
		A4K	clockwise	1.13
		A9K	back and forth	1.13
12	UD	A10	-	0.99
	CR	A11	back and forth	1.10

Samples were cut along the longitudinal sections of the final sheets and prepared for optical microscope examinations with the use of Barker etchant. Optical microscope images were obtained with a Zeiss M1m microscope using polarized light.

Samples with a diameter of 30 mm were cut for texture measurements and samples with 50 mm were cut for deep-drawing tests. Texture examinations were carried out with a Bruker D8 Advance X-ray diffractometer using CuKα radiation equipped with an Eulerian cradle operating with a 40 kV tube voltage and a 40 mA heating current. {111}, {200}, and {220} pole figures were measured up to CHI = 75° sample tilting. Complete recalculated {200} pole figures were obtained with the software of the equipment, TexEval. Orthotropic deformation was used during recalculation for both UD and CR samples.

The applied earing prediction method is described in detail elsewhere [12]. Briefly, recalculated and complete {200} pole figures were obtained by texture analysis from the measured {111}, {200}, and {220} incomplete pole figures. Then, the CHI-cuts, which show the recalculated intensity values versus CHI of the complete {200} pole figures were plotted for each PHI (sample rotation) angle.

The method applies the principle according to which the CHI-cuts can be approximated with a sum of Gaussian curves [14]. Thus, the CHI-cuts of the recalculated {200} pole figures were fitted with Gaussian curves. The area of each individual Gaussian curve was determined and then the areas were weighted (multiplied) with the sin of the CHI values corresponding to the peaks of the Gaussian curves. Afterwards, the weighted areas of the Gaussian curves were summarized to the given PHI. The summarized, weighted {200} intensities were plotted versus PHI, which shows the predicted relative cup height.

Deep-drawing tests were performed at the John von Neumann University with Erichsen deep-drawing equipment. The diameter of the samples was 50 mm and that of the drawn cups was 33 mm, giving a drawing ratio of 1.51. The applied blank holder force was 1 kN. Average earing was calculated as the ratio of average ear height and average cup height [12]. Predicted average earing was calculated similarly from the weighted {200} intensity curves and divided by a scaling factor. Since predicted average earing is calculated from measured X-ray intensity and not cup height, the use of a scaling factor is required to obtain comparable values. The determination of the scaling factor is based on comparison to experimental data. The scaling factor was determined by dividing the predicted average earing calculated form weighted {200} intensities with average earing. It was determined to both UD and CR cases. The obtained values were 40 and 60, respectively.

3. Results

3.1. Results of 6 Pass Rolling

Figure 1 depicts the optical micrographs of one UD and CR sample. It can be seen that the both rolling types resulted in elongated grains and notable differences cannot be observed.

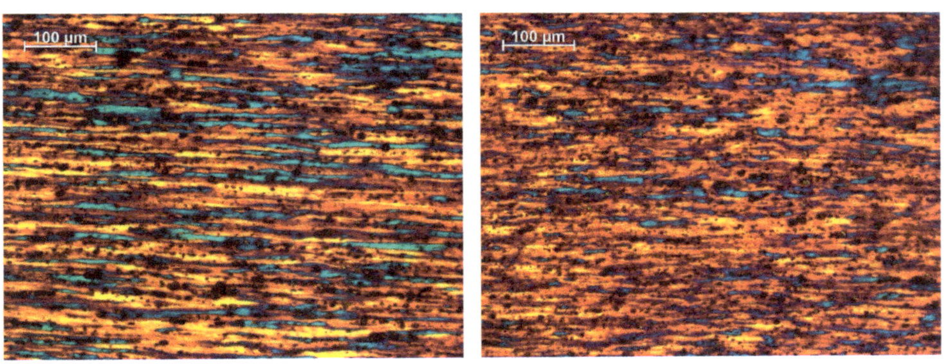

(**a**) (**b**)

Figure 1. Optical microscope images of the (**a**) UD (A5) and (**b**) CR (A4K) samples.

Figure 2 shows the recalculated {200} pole figures of the UD and CR samples rolled in 6 passes. It can be seen that the pole figures of the UD samples (Figure 2a,c,e) exhibit two-fold symmetry, while those of the CR samples (Figure 2b,d,f) exhibit four-fold symmetry.

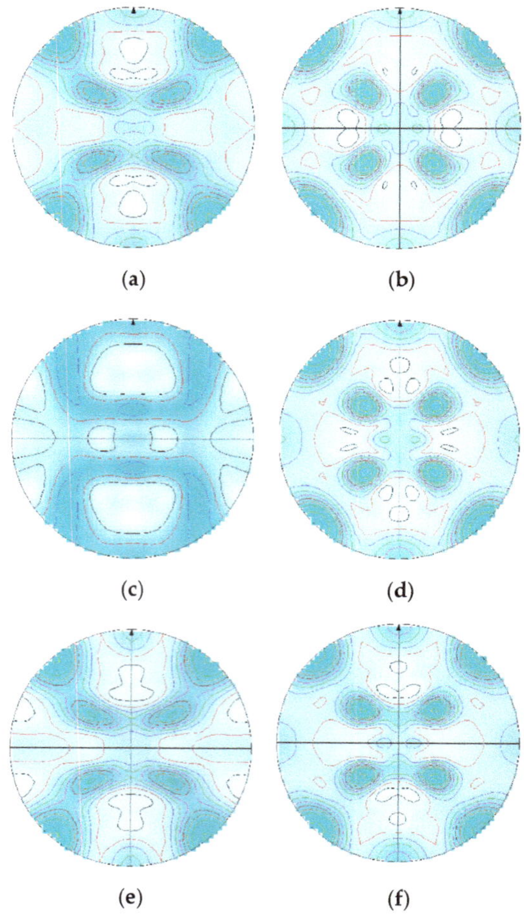

Figure 2. Recalculated {200} pole figures of the samples cold-rolled in 6 steps. (**a**) A1; (**b**) A3K; (**c**) A5; (**d**) A4K; (**e**) A6; (**f**) A9K.

In Figure 3, the cup heights of the samples rolled in 6 passes are summarized. Figure 3a shows the predicted relative cup heights (weighted {200} intensities) versus PHI for the UD samples and Figure 3b for the CR samples. Figure 3c,d show measured cup heights for the UD and CR samples, respectively. For both UD and CR samples, ears appear in the 45° + (n × 90°) directions on the predicted and measured cup height diagrams as well. Accordingly, rolling earing is observed after both UD and CR. It is evident that the positions of ears can be recognized more easily on the predicted relative cup height diagrams. However, based on these diagrams, it is quite difficult to tell which rolling mode resulted in larger earing. It is also worth noting that on the predicted relative cup height diagrams, small peaks are present in the 0° + (n × 90°) directions as well besides the ears. This was observed in a previous work as well, and was used to follow the evolution of rolling earing to recrystallized earing [12].

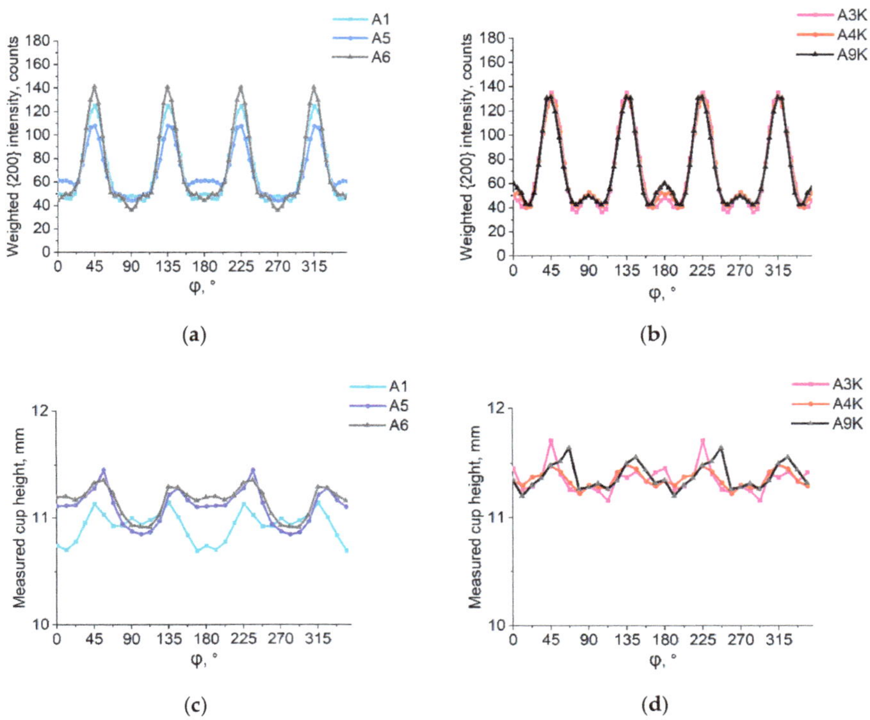

Figure 3. Cup heights of the samples cold-rolled in 6 steps. (**a**) Predicted relative cup heights of unidirectionally rolled samples; (**b**) predicted relative cup heights of cross-rolled samples; (**c**) measured cup heights of unidirectionally rolled samples; (**d**) measured cup heights of cross-rolled samples.

Average earing calculated from measured cup heights and predicted average earing calculated from weighted {200} intensities of UD and CR samples are plotted in Figure 4a,b, respectively. Note that the used scaling factor for UD was 40 and for CR it was 60. For convenience, the negative sign, which usually corresponds to rolling type earing, is neglected. It can be seen that the average earing values of the UD samples are around 2.5, while those of the CR samples are around 1.5. Thus, it can be stated that cross-rolling results in notably smaller earing compared to unidirectional rolling.

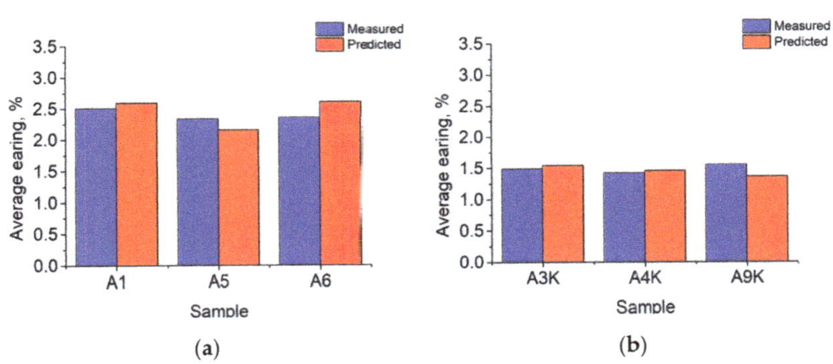

Figure 4. Absolute value of average earing of the samples cold-rolled in 6 steps. (**a**) Measured and predicted average earing of unidirectionally rolled samples, scaling factor: 40; (**b**) measured and predicted average earing of cross-rolled samples, scaling factor: 60.

3.2. Results of 12 Pass Rolling

In Figure 5, the recalculated {200} pole figures of the UD and CR samples rolled in 12 passes is presented. The two-fold symmetry of the UD sample (Figure 5a) and the four-fold symmetry of the CR sample (Figure 5b) are clearly seen.

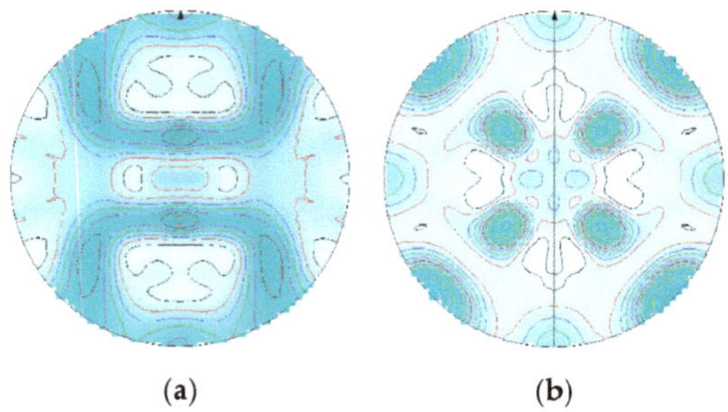

Figure 5. Recalculated {200} pole figures of the samples cold-rolled in 12 steps. (**a**) A10; (**b**) A11K.

Figure 6 shows cup heights of the samples rolled in 12 passes. In Figure 6a, the predicted relative cup height for the UD sample can be seen, and in Figure 6b that of the CR sample. Figure 6c,d plots measured cup heights for the UD and CR samples, respectively. Again, ears appear in the 45° + (n × 90°) directions on the predicted and measured cup height diagrams after both UD and CR, showing that rolling earing has been developed. The magnitude of earing is difficult to read from the predicted relative cup height and also from the measured cup height diagrams.

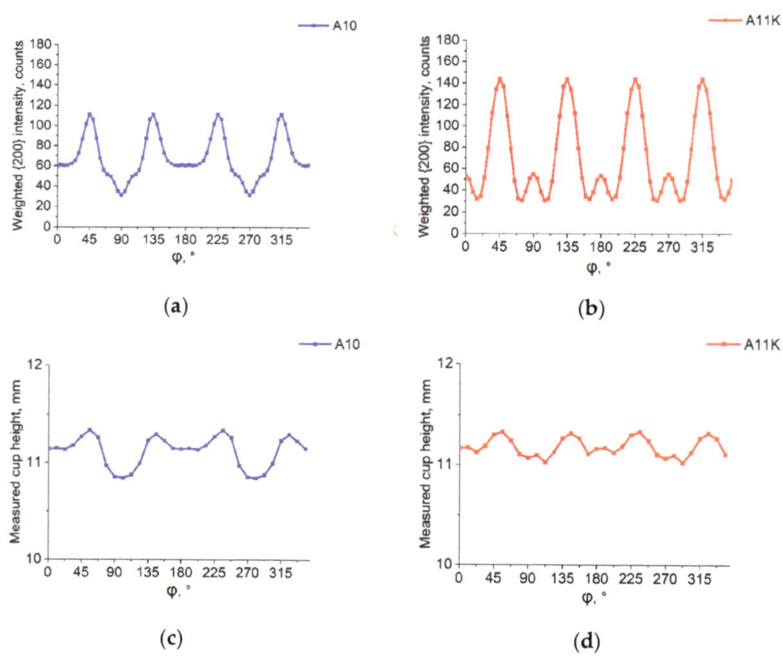

Figure 6. Cup heights of the samples cold-rolled in 12 steps. (**a**) Predicted relative cup height of the unidirectionally rolled sample; (**b**) predicted relative cup height of the cross-rolled sample; (**c**) measured cup height of the unidirectionally rolled sample; (**d**) measured cup height of the cross-rolled sample.

Average earing of UD and CR samples is plotted in Figure 7a,b, respectively. The values are around 2.2 for the UD and around 1.5 for the CR samples. It can be said that cross-rolled samples show smaller earing compared to unidirectionally rolled ones after 12 passes as well.

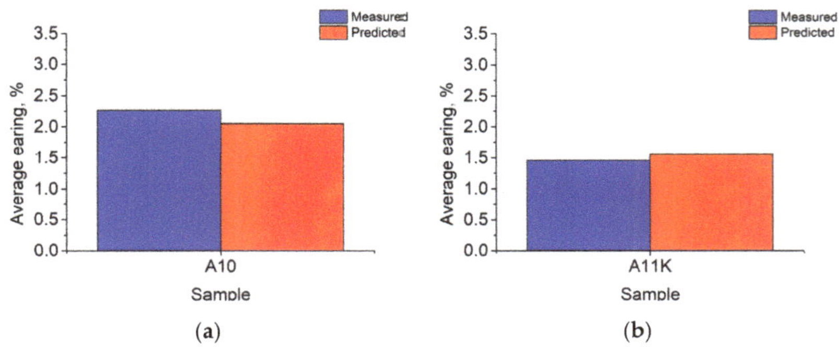

Figure 7. Absolute value of average earing of the samples cold-rolled in 12 steps. (**a**) Measured and predicted average earing of the unidirectionally rolled sample, scaling factor: 40; (**b**) measured and predicted average earing of the cross-rolled sample, scaling factor: 60.

4. Discussion

It was seen that after both 6 and 12 passes, the cross-rolled, 5056 type aluminum sheets exhibited more uniform, four-folded {200} pole figures compared to unidirectionally rolled sheets. In accordance, a notably smaller value of rolling-type earing (~1.5) was observed after cross-rolling compared to the unidirectionally rolled sheets (~2.2–2.5). As a consequence, a different scaling factor was required to predict average earing in the case of cross-rolling (60 instead of 40) even if all other parameters (including the parameters of cold rolling, texture measurements, deep-drawing, blank geometry, etc.) are unchanged. Finally, it can be concluded that besides unidirectional rolling, the developed method can be applied for cross-rolling as well.

Author Contributions: Conceptualization, M.B.; methodology, M.B.; validation, A.H. and B.S.; formal analysis, M.B.; investigation, M.B., A.H. and B.S.; data curation, B.S. and A.H.; writing—original draft preparation, M.B.; writing—review and editing, A.H.; visualization, B.S.; supervision, V.M.; funding acquisition, M.B. All authors have read and agreed to the published version of the manuscript.

Funding: Supported by the ÚNKP-19-4 New National Excellence Program of the Ministry for Innovation and Technology and the NKFIH K119566 project.

Acknowledgments: The authors are grateful for Gabor Beres from the John von Neumann University for his aid in carrying out the deep-drawing tests.

Conflicts of Interest: The authors declare no conflict of interest.

References

1. Neugebauer, R.; Glass, R.; Kolbe, M.; Hoffmann, M. Optimisation of processing routes for cross rolling and spin extrusion. *J. Mater. Process. Tech.* **2002**, *125–126*, 856–862. [CrossRef]
2. Li, X.; Al-Samman, T. Mechanical properties and anisotropy of ME20 magnesium sheet produced by unidirectional and cross rolling. *Mater. Des.* **2011**, *32*, 4385–4393. [CrossRef]
3. Wronski, S.; Wrobel, M.; Baczmanski, A.; Wierzbanowski, K. Effects of cross-rolling on residual stress, texture and plastic anisotropy in f.c.c. and b.c.c. metals. *Mater. Charact.* **2013**, *77*, 116–126. [CrossRef]
4. Huh, M.Y.; Cho, S.Y.; Engler, O. Randomization of the annealing texture in aluminum 5182 sheet by cross-rolling. *Mat. Sci. Eng. A* **2001**, *A315*, 35–46. [CrossRef]
5. Tang, W.; Huang, S.; Li, D.; Peng, Y. Mechanical anisotropy and deep drawing behaviors of AZ31 magnesium alloy sheets produced by unidirectional and cross rolling. *J. Mater. Process. Tech.* **2015**, *215*, 320–326. [CrossRef]

6. Fukui, S.; Kudo, H. The earing in deep-drawing and directionality in tension-test of sheet metal. *Rep. Inst. Sci. Tech. Univ. Tokyo* **1950**, *4*, 33.
7. Sowerby, R.; Johnson, W. Prediction of earing in cups drawn from Anisotropic sheet using slip-line field theory. *J. Strain Anal.* **1974**, *9*, 102–108. [CrossRef]
8. Tucker, G.E.G. Texture and earing in deep drawing of aluminium. *Acta Metall.* **1961**, *9*, 275–286. [CrossRef]
9. Kanatake, N.; Tozawa, Y.; Otani, T. Calculations from texture of earing in deep drawing for fcc metal sheets. *Int. J. Mech. Sci.* **1983**, *25*, 337–345. [CrossRef]
10. Van Houtte, P.; Cauwenberg, G.; Aernoudt, E. Analysis of the earing behaviour of aluminium 3004 alloys by means of a simple model based on yield loci calculated from orientation distribution functions. *Mater. Sci. Eng.* **1987**, *95*, 115–124. [CrossRef]
11. Engler, O.; Mertens, N.; Van Dam, P. Texture-based design of a convoluted cut-edge for earing-free beverage cans. *J. Mater. Process. Tech.* **2011**, *211*, 1278–1284. [CrossRef]
12. Benke, M.; Hlavacs, A.; Imre, P.; Mertinger, V. Prediction of earing of aluminium sheets from {h00} pole figures. *Eur. J. Mech. A Solid* **2020**, *81*, 103950. [CrossRef]
13. Sepsi, M.; Mertinger, V.; Benke, M. Sample cutting-free pole figure measurement method for centreless diffractometers in modified X mode. *Mater. Char.* **2019**, *151*, 351–357. [CrossRef]
14. Ma, Q.; Mao, W.; Feng, H.; Yu, Y. Rapid texture measurement of cold-rolled aluminum sheet by X-ray diffraction. *Scr. Mater.* **2006**, *54*, 1901–1905. [CrossRef]

metals

MDPI

Article

Physical and Structural Characterization of Monocrystalline Cu-13.7% Al-4.2% Ni Alloy Submitted to Thermo-Cyclical Treatments under Applied Loads

Lioudmila A. Matlakhova [1,*], Elaine C. Pereira [1], Serguey A. Pulnev [2], Carlos Y. Shigue [3] and Natalia A. Palii [4]

1. Advanced Materials Laboratory, Science and Technology Center, Darcy Ribeiro State University of Northern Rio de Janeiro, Campos dos Goytacazes 28013-620, Brazil; elainecp@pq.uenf.br
2. Division of Solid State Physics, Ioffe Institute, St. Petersburg 194021, Russia; pulnev@mail.ru
3. Lorena School of Engineering, University of São Paulo, São Paulo 05508060, Brazil; cyshigue@usp.br
4. Laboratory of Physical Chemistry of Amorphous and Nanocrystalline Alloys, A.A. Baikov Institute of Metallurgy and Materials Science, Russian Academy of Sciences, Moscow 119334, Russia; palii.natalia59@gmail.com

* Correspondence: lioudmila@uenf.br; Tel.: +55-229-8811-9042

Received: 19 December 2019; Accepted: 25 January 2020; Published: 4 February 2020

Abstract: Monocrystalline alloy with a nominal composition of Cu-13.7% Al-4.2% Ni (wt.%) that shows reversible martensitic transformations (RMTs) was studied. The alloy, manufactured by the "Memory Crystals Group" in Russia, was subjected to thermo-cyclical treatment (TCT) under tension within a range that included critical RMT temperatures. A special device was developed to perform TCTs (up to 500 cycles) and three different loads were applied: 0.11, 0.26, and 0.53 MPa. X-ray diffraction analysis, optical microscopy, differential calorimetry, and Vickers microhardness were involved in the alloy's characterization. Under TCTs, the alloy displayed complex structural transformation, revealing the sequence of RMT, $\beta_1 \leftrightarrow R \leftrightarrow \beta'_1 + \gamma'_1$; the involved phases were coherently precipitated but very sensitive to the experimental conditions. It was found that during TCTs (from 300 cycles) performed under optimum load (0.26 MPa), the processes of martensite reorientation, hardening, and stabilization of the structure were the most intensive thus leading to a reduction of RMT critical intervals and increased microhardness.

Keywords: Cu–Al–Ni monocrystalline alloy; reversible martensitic transformations; thermo-cyclic treatment under load; physical characterization and structural characterization

1. Introduction

Metal alloys exhibiting the shape memory effect (SME) have specific properties due to the fact of reversible martensitic transformations (RMTs) [1–3]. Widely used in industrial and medical applications, TiNi-based alloys have superb SME characteristics [3–6], but owing to their high cost, other shape memory alloys (SMAs) are under consideration. Copper-based SMAs cost less; moreover, they have high transformation temperatures, superior thermal and electrical conductivity, and large recoverable strain [7–9].

Monocrystalline Cu–Al–Ni (Cu_3Al-based) alloys have received special attention, because they display high SME parameters along with good thermoelastic properties, formability, and plasticity without embrittlement at high temperatures [7–13]. Such SMAs are based on the Cu_3Al intermetallic compound; they have different structures, depending on the composition, temperature, heat treatment, and deformation. At elevated temperatures, the Cu_3Al-based alloys have a structure of a β-phase, a body-centered cubic (BCC) phase, but below 565 °C, it undergoes the ordering to form the metastable β_1 phase, an ordered cubic structure of the DO_3 type [14,15]. With subsequent cooling and depending on the aluminum content, the ordered β_1 phase, in turn, can transform into either martensite γ'_1, the ordered phase of Cu_3Ti type, or martensite β'_1, the ordered phase of Cu_3Al type [1,3,15]. Aside from the aforementioned phases, the diffractograms of monocrystalline Cu–Al–Ni alloys may display additional peaks that can be attributed to the rhombohedral ordered Al_7Cu_4Ni phase (JCPDS: 28-0016) with the space-group R-3m of the trigonal crystal system [16,17]. In our previous work [12,18,19], it was shown that the microstructures of Cu–Al–Ni alloys revealed the presence of the rhombohedral ordered Al_7Cu_4Ni phase, so the additional peaks in diffractograms would be due to (1) the ordering of the Ni atoms incorporated into the crystalline lattice of alloy and/or (2) the development of a transitional crystal lattice state during RMT, $\beta_1 \leftrightarrow \gamma'_1$ [15–19]. The Al_7Cu_4Ni phase is called the R-phase in the present study as in our previous works [18–22].

Multiple heating/cooling cycles within a temperature range, which includes critical RMT temperatures, are required to perform a repetitive SME regime for investigated alloys. The restrictions in the practical application of SMAs are the possible undesirable changes in the SMAs' structure and performance after multiple thermal cycles. As is well known, thermos-cyclical treatments (TCTs) promote changes in the structure and phase composition of the alloy thus modifying its physical and mechanical properties [3,7–11]. In fact, each thermal cycle may produce additional imperfections, resulting in changes in RMT behavior, namely, the change in RMT critical temperatures [7,8,11].

Reversible martensitic transformations can develop under applied loads as well as thermal stresses in alloys that are used in force elements of several actuators. The applied loads give rise to tensions, deformations, and transformations of metastable phases (induced transformations) thus changing the structure and properties of the material in a complex way [4,7–13].

This work was conducted as the continuation of our work examining the effect of cyclic tests on monocrystalline alloys of the Cu–Al–Ni system on the structure, critical temperatures, and hysteresis of reversible martensitic transformations, involving multiple cycles of heating above A_f and cooling below M_f; we changed the test conditions and the composition of the studied alloys. So, in Reference [12], monocrystalline Cu-13.5Al-4Ni alloys were subjected to 100, 200, and 300 thermal cycles (TC) without load. In Reference [13], we studied two monocrystalline alloys: Cu-13,5Al-4Ni and Cu-13,7Al-4,2Ni in the initial state and after 1, 100, and 200 thermal cycles (TCs) without load. In Reference [18], monocrystalline Cu-13.5Al-4Ni alloy was subjected to thermal cycling treatments conducted for 100, 200, and 300 cycles free of load and under a static load of 0.5 kg, corresponding to a compressive stress of 0.28 MPa. This stress was a fraction of the alloy's "pseudo-yield" stress, determined as 60–65 MPa. In other tests, thermal cycling treatment was performed under different loads of 0.2 and 0.5 kg involving 300 cycles. In Reference [19], the same alloy was subjected to 300 thermal cycling treatment in a load-free regime and under three different applied load of 0.1, 0.2, and 0.5 kg, corresponding to 0.056, 0.112, and 0.280 MPa, respectively; and in Reference [20], this alloy passed through three distinct thermal cycling treatments involving 100, 200, and 300 cycles under one static load of 0.5 kg (0.280 MPa). These tests revealed that the monocrystalline Cu-13.5wt.% Al-4.0wt.% Ni alloy suffered reversible martensitic transformation, $\gamma'_1 \leftrightarrow \beta_1$ associated with a thermal hysteresis of about 20 °C when subjected to thermal cycling treatments. The phase transformation enthalpies of endothermic (direct) and exothermic (reverse) processes in the initial state and in a first of load-free thermal cycles were approximately 6.8–7.4 J/g. The thermal hysteresis of the RTM promotes changes in the number of phases associated with the RMT. These changes depend on the way the treatment ends, either by heating or by cooling. After 300 TCs without load, displacement of the critical temperatures M_o, M_f, A_o, A_s as well as the peak temperatures A_e and M_e is observed towards values up to 10 °C lower, while M_s increases by more than 10 °C. This contributes to enlarging the temperature interval of the RMT occurrence. By contrast, no change occurred in the thermal hysteresis. After a load-free 300 TCs, if the end occurs by heating above A_f, a larger amount of β_1 is transformed. If the end occurs by cooling below M_f, an intermediate structure, the R-phase, related to the $\gamma'_1 \leftrightarrow \beta_1$ is found.

This is apparently a consequence of accumulated structural defects that retard the RMT and tend to stabilize the intermediate stage. The alloy submitted to TCs under a load of 0.5 kg revealed a tendency towards a decrease in the critical intervals for both direct and reverse martensite transformation with an increasing number of cycles. This tendency is accentuated for treatments up to 300 cycles in which critical temperature intervals of RMTs smaller than 10 °C were recorded. The hysteresis of the RMT also decreased. This can be attributed to a higher participation of the R-phase, particularly above 200 cycles, either alone or coherent with γ'_1 (direct on cooling) and β_1 (reverse on heating). It is also suggested that the applied load contributes to the shape deformation observed in blocks of martensite variants. Despite complex structural transformations, the monocrystalline Cu-13.5wt.% Al-4.0wt.% Ni alloy showed good resistance to irreversible changes during applied tests.

In a more recent work [21], we paid attention to the behavior of a monocrystalline Cu-13.7Al-4.2Ni alloy subjected to thermal cycling treatment induced for 1, 100, 200, 300, 400, and 500 cycles without applied load, i.e., stress free.

In the current work, which revisits and expands our research, monocrystalline Cu-13.7% Al-4.2% Ni alloy was subjected to up to 500 cycles under three constant loads within a critical RMT temperatures range along with the physical and structural characterization of the alloy during applied TCTs.

2. Materials and Methods

The basic material investigated was a high purity monocrystalline Cu-13.7wt.% Al-4.2wt.% Ni alloy, obtained as a cylindrical bar (4 mm in diameter) from the Memory Crystals Group of the Technical University of Saint Petersburg, Russia [23]. Specimens approximately 4 mm in thickness were sectioned out perpendicular to the bar axis by means of a Minitom cutter (Struers A/S Pederstrupvej, Denmark). For microstructure characterization, the specimens were polished to an acceptable mirror finish using 0.1 μm alumina paste. No chemical etching was used, since the surface relief of transformation was enough to reveal the phases.

Specimens were submitted to 1, 100, 200, 300, 400, and 500 thermal cycles of heating up to 90 °C (above A_f—in the initial state) followed by cooling to 0 °C (below M_f—in the initial state). Each TCT was carried under a static load (TCTL). Three loads were applied: 0.14, 0.35, and 0.70 kg, corresponding to the stress of 0.11, 0.26, and 0.53 MPa, respectively. The chosen loads were below the flow limit determined as 70 MPa for this alloy [18,21]. The load was applied along with thermocycling using the special device developed for this purpose [18–22]. After TCTL, the structure of the alloy was analyzed by optical metallography and X-ray diffraction. Structural characterization was conducted at room temperature (RT) by finishing the treatment: (a) with a half cooling cycle to 0 °C, then up to RT, and (b) with a half heating cycle to 90 °C, then down to RT [12,18–22].

The phase structure of the alloy was determined at RT by X-ray diffraction (XRD) in both Shimadzu 7000 (SHIMADZU Corporation, Kyoto, Japan) and Seifert 65 diffractometers (Seifert GmbH, Ahrensburg, Germany) using Cu-K_α and Co-K_α radiations, respectively, for 2θ angles from 25° to 75° at scanning steps of 0.03°/3 s.

The surface morphological relief induced by the phase transformation and the involved deformation-related defects were observed in a Neophot-32 microscope (Carl Zeiss AG, Oberkochen, Germany) with polarized light. The microhardness was measured with a Shimadzu HMV-2 Micro Vickers Hardness Tester (Shimadzu Corporation, Kyoto, Japan). The RMT critical temperatures and associated thermal effects after the TCTL were determined using a Q10 differential scanning calorimeter (TA Instrument Co. Ltd., New Castle, DE, USA). The thermal tests were conducted in the temperature range from −50 °C to +150 °C with a heating/cooling rate of 10 °C/min in an inert atmosphere. Cooling was carried out using an LNCA mode (Liquid nitrogen cooling). After TCTL, the samples were submitted to two thermal cycles during the DSC runs. In the first cycle of the DSC analysis, the samples were cooled to −50 °C, then heated to +150 °C, and cooled and heated again within the indicated temperature range. To determine the thermal hysteresis value in our studies, we made use of the difference between the DSC peaks' critical temperatures upon heating and cooling (ΔT = Ap − Mp), according to Morin and Trivero [11].

3. Results

3.1. Analysis of the Monocrystalline Cu-13.7% Al-4.2% Ni Alloy in the Initial State

Considering that the critical RMT temperatures of the studied alloy cover the ambient temperature (AT), special care was given to the structural analysis of the alloy before TCT. In other words, the structure of the alloy was examined in the as-received state and during the first load-free thermal cycle ending with the 1/2 cooling cycle (AT → 0 °C → AT), heating (AT → +90 °C → AT), and performing the XRD analysis at room temperature (Figure 1).

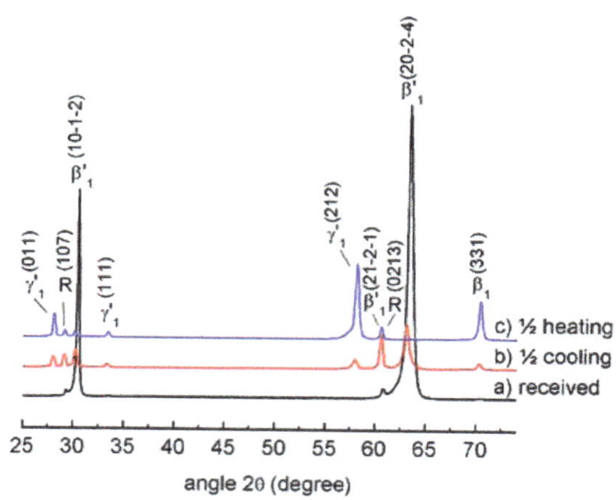

Figure 1. The XRD patterns of the alloy (**a**) in the as-received state, (**b**) after the 1/2 cooling cycle, and (**c**) after the 1/2 heating cycle.

The XRD pattern of the alloy in the as-received state in Figure 1a shows a significant amount of the β'_1 martensitic phase, determined by the intensive β'_1 and β'_1 peaks, traces of the R-phase, determined by the peak (107)R, and the other peak found for the R-phase coinciding with the peak belonging to the β'_1 phase and β'_1|(0213)R.

After the 1/2 cooling cycle, in addition to the major β'_1 phase peaks, β'_1 and β'_1, the XRD pattern displays the martensitic γ'_1 peaks, (011)γ'_1, (212)γ'_1, and (111)γ'_1, the R-phase, peaks, (107)R, β'_1|(0213)R, and the minor presence of the β_1 phase, peak $(331)_{\beta 1}$ (Figure 1b). The structure of the alloy submitted to 1/2 heating cycle (Figure 1c) varied deeply in the participation of the aforementioned phases: the greater the participation of the martensite γ'_1 phase, there was some increase in the participation of the high-temperature β_1 phase, (331)β_1, while the presence of the β'_1 and R-phases was less intensive.

The DSC analysis of monocrystalline Cu-13.7% Al-4.2% Ni alloy in the initial state (Figure 2) showed that on the second heating, the reverse RMT ($\beta'_1 + \gamma'_1$) → β_1 occurred in the range from +38.2 °C (A_s) to +68.8 °C (A_f) through an endothermic process with a phase transformation enthalpy of 9 J/g. The process developed an "on-set" mode from +51.4 °C (A_o), presenting the peak temperature of +54.7 °C (A_p), where the heat flux was maximum. The process ended in an "off-set" mode at the temperature of +60.4 °C (A_{of}). On cooling of the alloy, the direct RMT β_1 → ($\beta'_1 + \gamma'_1$) occurs between +51.4 °C (M_s) and 16.0 °C (M_f), having on-set temperature peaks 47.7 °C (M_o), +42 °C (M_p) and +32.6 °C (M_{of}), respectively, attributed to the exothermic process with a transformation enthalpy of 8.8 J/g.

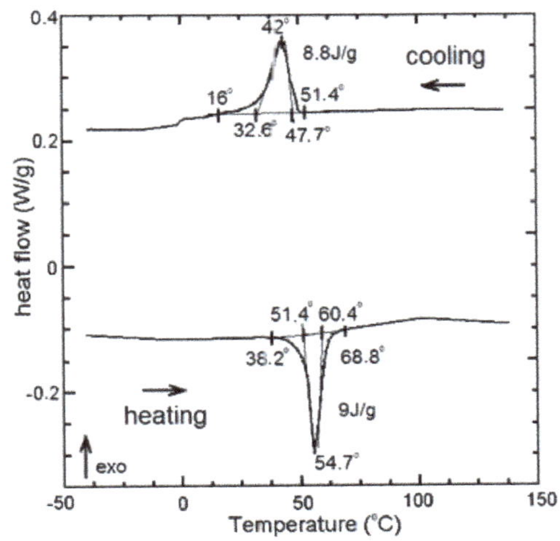

Figure 2. The DSC analysis of monocrystalline Cu-13.7% Al-4.2% Ni alloy in the initial state showed reverse heating and direct reversible martensitic transformations (RMT).

It is known that SME Cu-Al-Ni alloys can undergo a single transformation ($\beta_1 \rightarrow \beta'_1$ or $\beta_1 \rightarrow \gamma'_1$) or a mixed transformation ($\beta_1 \rightarrow \gamma'_1 + \beta'_1$) depending on the composition of the alloy. In [24–29] it was confirmed that the mixed transformation ($\beta_1 \rightarrow \gamma'_1 + \beta'_1$) takes place in the structure of Cu–Al–Ni alloys of approximately the same composition as the studied alloy. These martensitic transformations to γ'_1 or β'_1 structures have negligible energy difference so that sliding along the atomic layers can be implemented. Thus, the transformation from one compact structure to another is easily accomplished [3,30].

The DSC results show that during heating/cooling of the alloy, the critical RMT ranges were narrow, approximately 30.6 °C for A_s-A_f and 35.4 °C for M_s-M_f. As a measure of thermal hysteresis, the difference among the peak critical temperatures ($\Delta T = A_p - M_p$) was determined at approximately 12.7 °C in the first thermal cycle [20–22]. According to References [3,24–29], RMT features can be deduced from the magnitude of the thermal hysteresis which is in good agreement with the results obtained, whereas the martensitic transformation $\beta_1 \leftrightarrow (\beta'_1 + \gamma'_1)$ presented hysteresis around 10 °C (Figure 3).

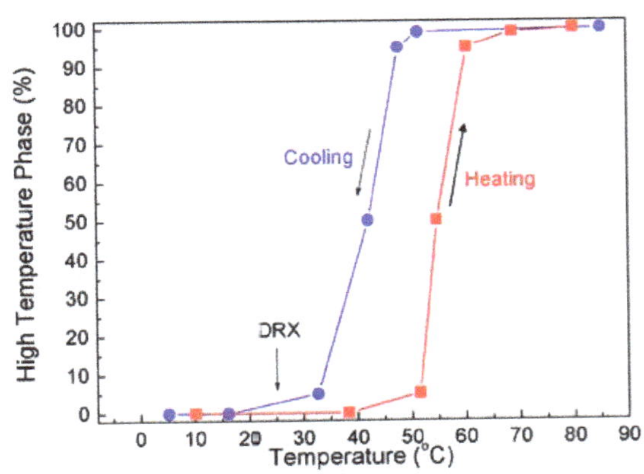

Figure 3. Schematic representation of the thermal hysteresis, showing the transformation $\beta_1 \leftrightarrow (\beta'_1 + \gamma'_1)$, presenting hysteresis around 10 °C.

3.2. *Differential Scanning Calorimetry of the Monocrystalline Cu-13.7% Al-4.2% Ni Alloy after TCT under Loads*

3.2.1. Reverse RMT Curves during the First Cycle of DSC Analysis

Thermal cycling-treated samples under loads were submitted to two thermal cycles in the DSC analysis. Firstly, the samples were cooled from RT. The reverse RMT curves obtained in the first DSC cycle of monocrystalline Cu-13.7% Al-4.2% Ni alloy submitted to TCT under load are shown in Figure 4. After TCT under loadings of 0.11 MPa (Figure 4a) and 0.53 MPa (Figure 4c), the reverse RMT peak was split into two. In contrast, at TCT under a load of 0.26 MPa, the reverse RMT curves of the alloy showed a single transformation peak; thus, structural imperfections accumulated during TCTs promote full RMT (Figure 4b).

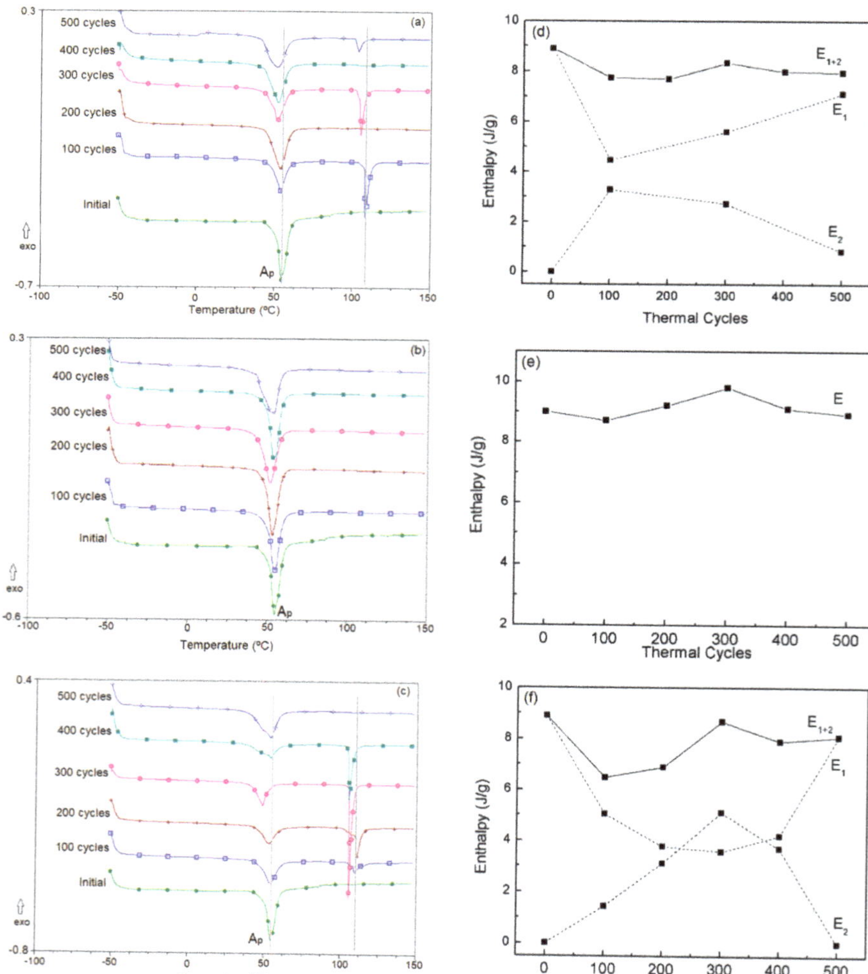

Figure 4. The reverse RMT curves (**a–c**) and reverse RMT enthalpies (E) (**d–f**) obtained in the first DSC cycle of monocrystalline Cu-13.7% Al-4.2% Ni alloy submitted to TCT under loadings of (**a**) 0.11 MPa, (**b**) 0.26MPa, and (**c**) 0.53 MPa.

This result becomes more evident when considering reverse RMT enthalpies (E) of the alloy after TCT under the loads of 0.11 (Figure 4d), 0.26 (Figure 4e), and 0.53 MPa (Figure 4f), where E_1 and E_2 are the enthalpy of the first and the second transformation peak, respectively; E_{1+2} is the sum of the enthalpies.

It is interesting to monitor the evolution of the enthalpies E_1 and E_2. The value of E_2 tends to increase with an increase in the number of cycles up to 100 TCT (at 0.11 MPa) or up to 300 TCT (at 0.53 MPa) and decrease with a further increase in the number of TCTs under load. The value of enthalpy was entire in the case of TCTs under 0.26 MPa (Figure 4e). The total enthalpy of the reverse transformation, in all tests, varied only slightly with the number of cycles.

An increase in the critical temperature range of complete transformation $(\beta'_1 + \gamma'_1) \rightarrow \beta_1$ correlate with the accumulation of lattice distortions during TMR/TCT, possible structural imperfections, its decrease with "training", and the predominant orientation of martensitic lamellas. Apparently, the load at 0.11 MPa was not large enough for "training" work, while the load at 0.53 MPa promoted the lattice distortions and the accumulation of its imperfections and complicated the implementation of TMR for the bulk material in the temperature range adopted in the TCT. In other words, under 0.11 MPa and 0.53 MPa, the TCT was realized in conditions of incomplete RMT cycles.

In these tests, a load at 0.26 MPa seemed to be sufficient to give rise to the reversible martensitic transformations during heating and cooling thus facilitating the "training" of the alloy during cyclic tests and maintaining a narrow range of TMR and did not stimulate the accumulation of defects in the alloy structure.

3.2.2. Direct and Reverse RMT of the Monocrystalline Cu-13.7% Al-4.2% Ni Alloy Curves after TCT under Loads

The DSC analysis of the alloy in the initial state showed that on heating, the reverse RMT $\beta'_1 \rightarrow \beta_1$ occurred in the range of +38.2 °C (A_s) to +68.8 °C (A_f) through an endothermic process: phase transformation enthalpy ~9 J/g. The process developed from an on-set temperature of +51.4 °C (A_o) and was responsible for the peak at +54.7 °C (A_p), where the heat flux was maximum. The process ended in an "off-set" mode at +60.4 °C (A_{of}). On cooling, the direct RMT, $\beta_1 \rightarrow \beta'_1$, occurred in the temperature range of +51.4 °C (M_s) to 16.0 °C (M_f), having an on-set, peak and off-set temperatures at +47.7 °C (M_o), +42 °C (M_p), and +32.6 °C (M_{of}), respectively, associated with the exothermic process with a transformation enthalpy of 8.8 J/g (Figure 2).

The effect of TCTs under the loads of 0.11, 0.26, and 0.53 MPa on the RMT characteristics of the monocrystalline Cu-13.7% Al-4.2% Ni alloy as determined by DSC analysis during cooling and 2nd heating without load and manifested in the displacement of critical temperatures and intervals is shown in Figures 5–7.

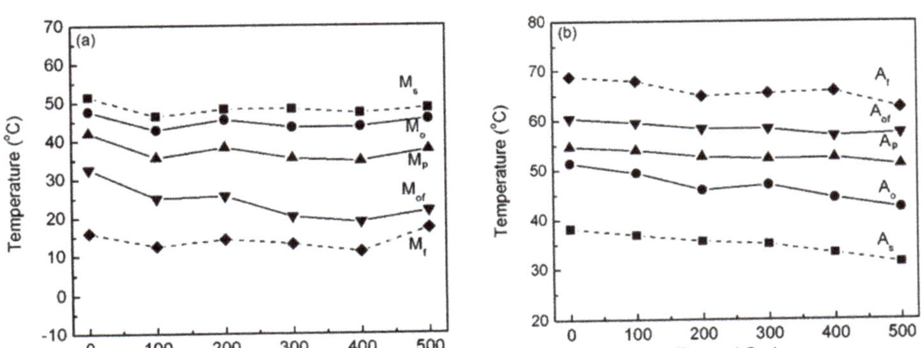

Figure 5. RMT critical temperatures obtained by DSC analysis during (**a**) cooling and (**b**) 2nd heating without loading of the monocrystalline Cu-13.7% Al-4.2% Ni alloy after TCT under the load of 0.11 MPa.

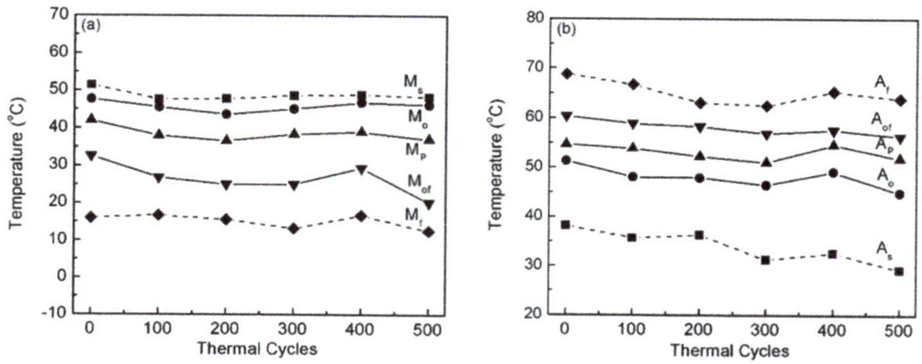

Figure 6. RMT critical temperatures obtained by DSC analysis during (**a**) cooling and (**b**) 2nd heating without loading of the monocrystalline Cu-13.7% Al-4.2% Ni alloy after TCT under the load of 0.26 MPa.

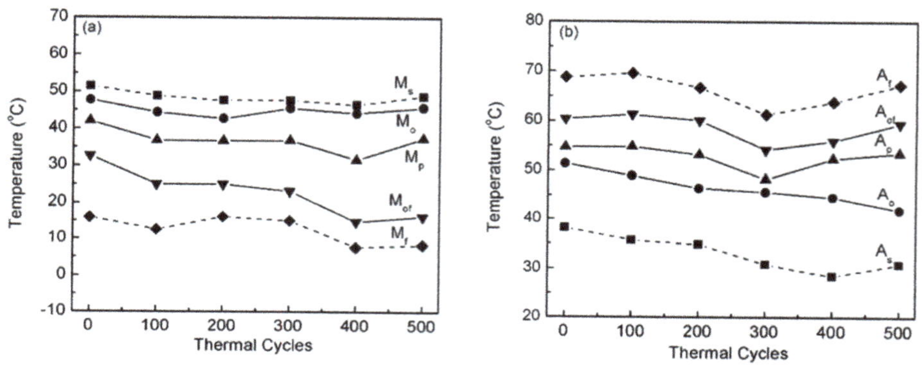

Figure 7. RMT critical temperatures obtained by DSC analysis during (**a**) cooling and (**b**) 2nd heating without loading of the monocrystalline Cu-13.7% Al-4.2% Ni alloy after TCT under the load of 0.53 MPa.

After TCTs under the load of 0.11 MPa, the critical temperatures of intensive RMT, mainly M_{of} and A_o, moved slightly to lower values, thus justifying the increase in the intensive RMT intervals. The intensive direct RMT ranges, M_o-M_p and M_o-M_{of}, increased from 5.7 and 15.1 °C (in the first cycle without loading) to 8.1 and 23.3 °C (300 TCTs) and remained almost constant at 8 and 23.9 °C (500 TCTs). The intensive intervals of reverse RMT, A_o-A_p and A_o-A_{of}, during the first thermal cycle, determined as 3.3 and 9 °C, increased to 8.7 and 15.1 °C, respectively, when it came to 500 TCTs (Figure 5).

After TCTs under the load of 0.26 MPa, the critical temperatures of intensive RMTs shifted slightly to lower values, up to 300 TCTs; then, the intensive temperatures remained constant and/or increased in value. The intensive direct RMT ranges, M_o-M_p and M_o-M_{of}, increased from 5.7 and 15.1 °C (in the first cycle without loading) to 9.1 and 26.1 °C (500 TCTs). The intensive intervals of reverse RMT, A_o-A_p and A_o-A_{of}, during the first thermal cycle were determined as 3.3 and 9 °C and increased to 7 and 11.5 °C (500 TCTs) (Figure 6).

After TCTs under a load of 0.53 MPa, the critical temperatures of intensive direct RMT, M_p and M_{of}, moved slightly to lower values, up to 400 TCTs, and remained the same. The intensive direct RMT intervals, M_o-M_p and M_o-M_{of}, increased from 5.7 and 15.1 °C (the first cycle without loading) to 8.4 and 29.7 °C after 500 TCTs. The critical temperatures of intensive reverse RMT, A_o, A_p, and A_{of}, moved slightly to lower values, up to 300 cycles, and tended to increase afterward. The intensive ranges of reverse RMT, A_o-A_p and A_o-A_{of}, during the first thermal cycle were determined as 3.3 and

9 °C, then decreased to 2.7 and 8.6 °C after 300 cycles, and finally increased to 11.7 and 17.5 °C with 500 cycles TCT (Figure 7).

The difference among the peak critical temperatures ($\Delta T = A_p - M_p$), as a measure of the thermal hysteresis amplitude, was determined as approximately 12.7 °C in the first thermal cycle of the investigated alloy. The changes in peak temperatures, A_p and M_p, after TCT under load, alter the magnitude of the thermal hysteresis as observed in Figure 8.

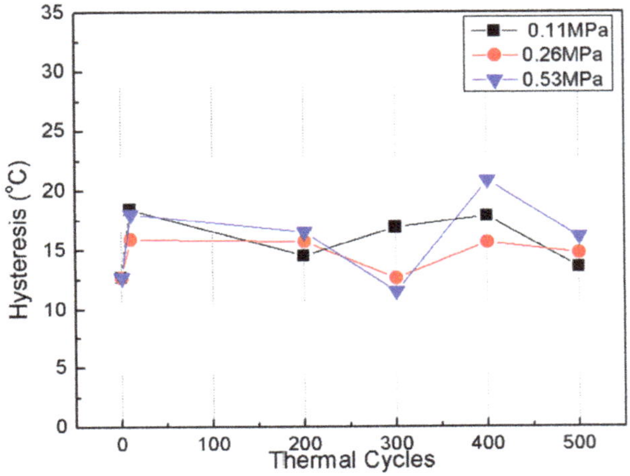

Figure 8. Values of thermal hysteresis as a function of applied TCT of the monocrystalline Cu-13.7% Al-4.2% Ni alloy.

3.3. *XRD Analysis of the Monocrystalline Cu-13.7% Al-4.2% Ni Alloy after TCT under Load*

3.3.1. XRD Analysis after TCT under Loading of 0.11 MPa Ending with the 1/2 Cooling Cycle

The Cu-13.7% Al-4.2% Ni alloy diffractograms after TCT, terminated with the 1/2 cooling cycle (under load) and the 1/2 heating cycle (without the load), are shown in Figure 9a,b. As can be seen, the phase compositions of these alloys differ significantly.

The diffractogram of Cu-13.7% Al-4.2% Ni alloy after TCTs under the load of 0.11 MPa finished with the 1/2 cooling cycle exhibited four metastable phases (Figure 9a). The major phase was martensite β'_1, an ordered $AlCu_3$-type phase [31], peaks β'_1 and β'_1. The other martensitic phase γ'_1, ordered Cu_3Ti-type phase [1,3,15] was identified with the peaks $(011)\gamma'_1$, $(212)\gamma'_1$, and $(111)\gamma'_1$. The high temperature phase β_1, ordered BiF_3-type phase [14] was detected by its $(331)_{\beta 1}$ peak. The Al_7Cu_4Ni phase with rhombohedral structure [16,17], called the R-phase in our works [18–22], is represented by the (107)R peak as well as the peak corresponding to the plane of coherence between the phases β'_1 and R, $\beta'_1|(0213)R$.

With the accumulation of 100 cycles, the major phase, martensite β'_1, was identified by intense β'_1 and β'_1 peaks; the moderate (107)R peak corresponded to the R-phase. A peak related to the plane of coherence between the martensitic β'_1 phase and R, $\beta'_1|(0213)R$ was detected. The other peaks, $(011)\gamma'_1$, $(111)\gamma'_1$, $(212)\gamma'_1$ and $(331)\beta_1$, had low intensity.

With the accumulation of 200 thermal cycles, the diffractogram shows minor changes in structure, reducing only the intensity of the coherence planes peak $\beta'_1|(0213)R$ and the R-phase peak (107)R. The martensitic β'_1 phase was still observed, indicating its temperature/thermal stability. After 300 TCTs, the accumulated distortions promoted changes in the martensitic structure, and the induced martensitic transformation $\beta'_1 \rightarrow \gamma'_1$ was observed. This TCT treatment promotes the greater stability of the martensitic phase γ'_1, determined by (011)γ'_1 and (212)γ'_1 peaks. The plane of coherence between the martensitic phases β'_1 and R, β'_1|(0213)R, display moderate intensity.

The treatment of 400 TCTs once again results in major participation of martensite phase β'_1 determined through the β'_1 - peaks. After 500 TCTs the XRD patterns reveal intense coherent planes β'_1|(0213)R peak, as well as β'_1 and β'_1 peaks, corresponding to martensitic β'_1. The (107)R peak, referring to phase R, has moderate intensity. The other peaks, (011)γ'_1, (111)γ'_1, (212)γ'_1 and (331)β_1, have low intensity.

3.3.2. XRD Analysis after TCT under Loading of 0.11 MPa Ending with 1/2 Heating Cycle

With the accumulation of 100 cycles, the major phase is martensitic β'_1, represented by (10-1-2)β'_1 and (20-2-4)β'_1 peaks. The peaks of coherent plane, β'_1|(0213)R, of the R-phase, (107)R, and of the high temperature phase β_1, (331)β_1, display moderate intensity. The martensitic phase γ'_1 is detected by its low intensity peaks (011)γ'_1, (111)γ'_1, and (212)γ'_1, Figure 9b.

With the accumulation of 200 TCTs, the amount of martensite β'_1 decreased and martensite γ'_1 increased, exhibiting $(011)_{\gamma'1}$ and $(212)_{\gamma'1}$ peaks. The peak of the coherent plane β'_1|(0213)R was intensive, whereas the R-phase peak, (107)R, exhibited moderate intensity.

After 300 TCTs, the XRD pattern changed slightly, displaying higher participation of the martensitic β'_1 phase, and the (10-1-2)β'_1 and (20-2-4)β'_1 peaks reappeared.

After 200 and 300 TCTs, the induced martensitic transformation $\beta'_1 \rightarrow \gamma'_1$ was incomplete (Figure 9b). The energy difference among these martensitic structures was negligible, and the sliding along of the atomic layers could be implemented. Thus, the transformation from one close-packed structure to the other was easily performed, mostly during thermal cycling [3,30].

The 400 TCTs induced considerable participation of the coherent plane β'_1|(0213)R and of the martensitic β'_1 (10-1-2)β'_1 and (20-2-4)β'_1 peaks. Reorientation of the high-temperature phase β_1 was revealed by an intensive (400)β_1 peak. The peaks associated with the γ'_1 and R-phases, (011)γ'_1, (212)γ'_1, and (107)R were less intensive.

After 500 TCTs, the XRD pattern displayed a higher amount of the martensitic β'_1 phase, represented by (10-1-2)β'_1 and (20-2-4)β'_1 peaks, a characteristic peak of the R-phase, (107)R, but small participation of the coherent plane β'_1|(0213)R.

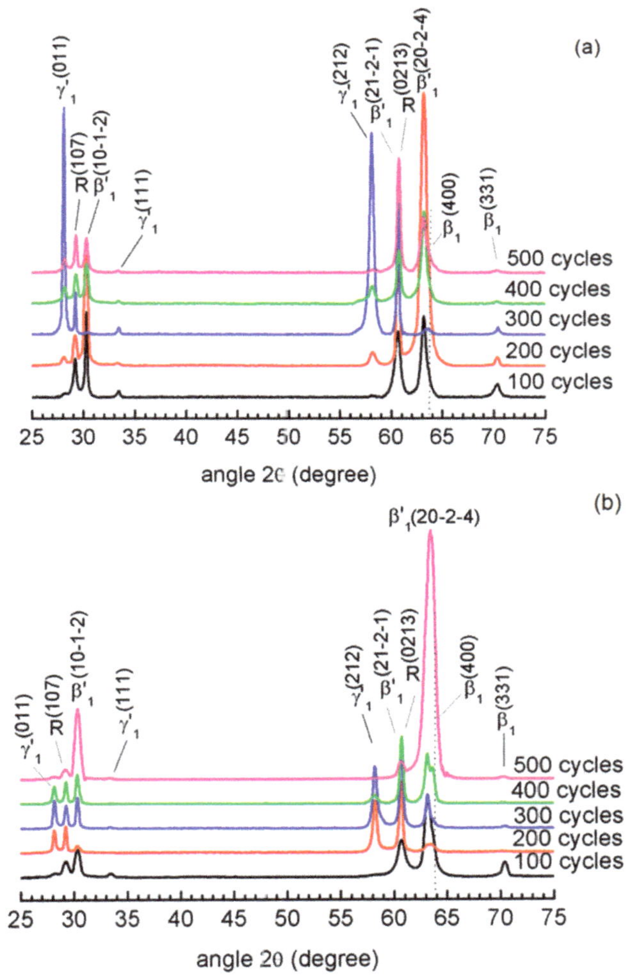

Figure 9. Diffractograms of Cu-13.7% Al-4.2% Ni alloy after TCTs under loading of 0.11 MPa finished with the (**a**) 1/2 cooling cycle and (**b**) 1/2 heating cycle.

3.3.3. XRD Analysis after TCT under Loading of 0.26 MPa Ending with 1/2 Cooling Cycle

Figure 10a shows XRD patterns of Cu-13.7% Al-4.2% Ni alloy after TCTs under the load of 0.26 MPa ending with the 1/2 cooling cycle (under load). After 100 TCTs, the pattern exhibited a major peak corresponding to the β_1 phase, $(331)\beta_1$; the peak corresponding to the coherent planes β'_1 with R, β'_1/(0213)R, was detected with moderate intensity, while peaks of the martensitic γ'_1, $(111)_{\gamma'1}$, and of the R-phase, $(107)_R$, had lower intensity.

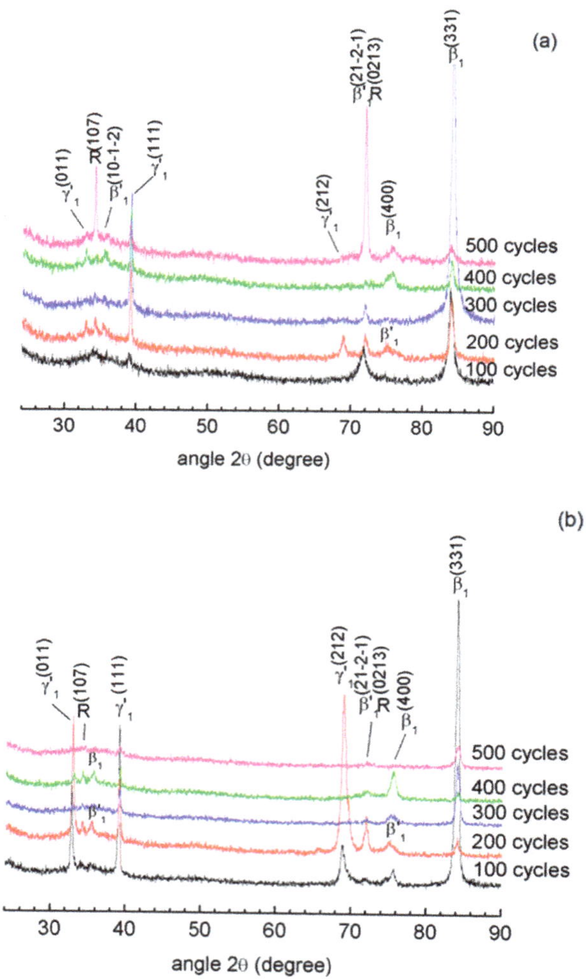

Figure 10. Diffractograms of Cu-13.7% Al-4.2% Ni alloy after TCTs under loading of 0.26 MPa finished with the (**a**) 1/2 cooling cycle and (**b**) 1/2 heating cycle.

With the accumulation of 200 TCTs, considerable participation by the martensite γ'_1, (111)γ'_1, (011)γ'_1, and (212)γ'_1 peaks and of the high-temperature phase β_1, (331)β_1 peak, was observed. The β'_1 martensite, with the less intensive (10-1-2)β'_1 and (20-2-4)β'_1 peaks, as well as the R-phase, (107)R, and the coherent plane, $\beta'_1|$(0213)R, were also detected. After 300 TCTs, the high-temperature β_1 phase displayed an intensive (331)β_1 peak. The martensite γ'_1 was detected by a moderate intensity (111)γ'_1 peak, while the peak corresponding to the coherent plane between the phases β'_1 and R, $\beta'_1|$(0213)R, had lower intensity.

With the accumulation of 400 TCTs, the XRD pattern indicated greater participation by the martensite γ'_1 and high-temperature β_1 phase as represented by (111)γ'_1, (011)γ'_1, (200)β_1, (400)β_1, and (331)β_1 peaks. The R-phase peak, (107)R, had a small intensity. After 500 TCTs, the XRD pattern revealed the R-phase, as a major phase, represented by intensive (107)R and $\beta'_1|$(0213)R peaks. The other (011)γ'_1, (018)R, (111)γ'_1, (400)β_1, and (331)β_1 peaks had lower intensity.

3.3.4. XRD Analysis after TCT under Loading of 0.26 MPa Ending with 1/2 Heating Cycle

The Cu-13.7% Al-4.2% Ni alloy XRD patterns after TCTs under 0.26 MPa load finished with the 1/2 heating cycle are shown in Figure 10b. With the accumulation of 100 cycles, high-temperature phase β_1 displayed (331)β_1 and (400)β_1 peaks, and the martensitic γ'_1 phase was detected by its (011)γ'_1, (111)γ'_1, and (212)γ'_1 peaks.

The accumulation of 200 TCTs favored the greater participation by martensite γ'_1, detected by (111)γ'_1, (011)γ'_1, and (212)γ'_1 peaks. The other phases displayed (107)R, (10-1-2)β'_1, (20-2-4)β'_1 β'_1|(0213)R, and (331)β_1 peaks with lower intensity. After 300 TCTs, the considerable participation of the high-temperature β_1 phase revealed by (331)β_1 and (400)β_1 peaks were observed. The martensite γ'_1 was detected by its moderate intensity peak. (111)γ'_1, while the phase R peaks, (107)R, and the coherent plane between the phases β'_1|R, β'_1|(0213)R, showed low intensity.

Accumulation of 400 TCTs resulted in the greater participation by the high-temperature phase β_1 as represented by (200)β_1, (400)β_1, and (331)β_1 peaks. The martensite γ'_1 displayed moderate intensity (011)γ'_1 and (111)γ'_1 peaks. The phase R peak, (107)R, and the plane of coherence between phases β'_1|R, β'_1|(0213)R, were revealed by lower intensity.

After 500 TCTs, the XRD pattern indicated the high-temperature phase β_1 as a major phase, identified with (331)β_1 peak. Low-intensity (107)R, (111)γ'_1, and β'_1|(0213)R peaks corresponded to minor phases.

3.3.5. XRD Analysis after TCT under Loading of 0.53 MPa Ending with 1/2 Cooling Cycle

Cu-13.7% Al-4.2% Ni alloy diffractograms after TCT under the load of 0.53 MPa finished with the 1/2 cooling cycle are shown in Figure 11a. With the accumulation of 100 TCTs, the prominent participation of the martensite β'_1 was represented by high (10-1-2)β'_1 and (20-2-4)β'_1 peaks as well as by the coherent plane β'_1|(0213)R between β'_1 and R-phases and the (107)R peak of the R-phase. The (011)γ'_1, (111)γ'_1, and (212)γ'_1 peaks of martensite γ'_1 and the (331)β_1 peak of the high-temperature β_1 phase showed moderate intensity, representing smaller participation by these phases. With the accumulation of 200 TCTs, the coherent plane β'_1|(0213)R peak became higher, displaying the greater participation of β'_1|R. The participation of the martensitic γ'_1 and β'_1 phases, as well as R-phase, were detected by the moderate intensity peaks: (011)γ'_1, (111)γ'_1, (212)γ'_1, (10-1-2)β'_1, (20-2-4)β'_1, and (107)R. The (331)β_1 peak of the high-temperature β_1 phase had the lowest intensity.

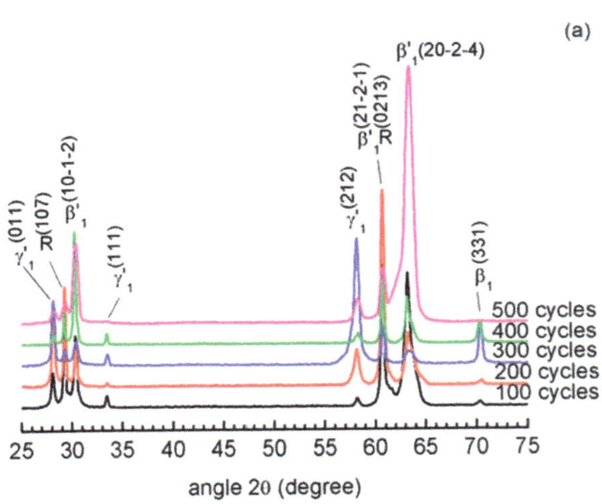

Figure 11. *Cont.*

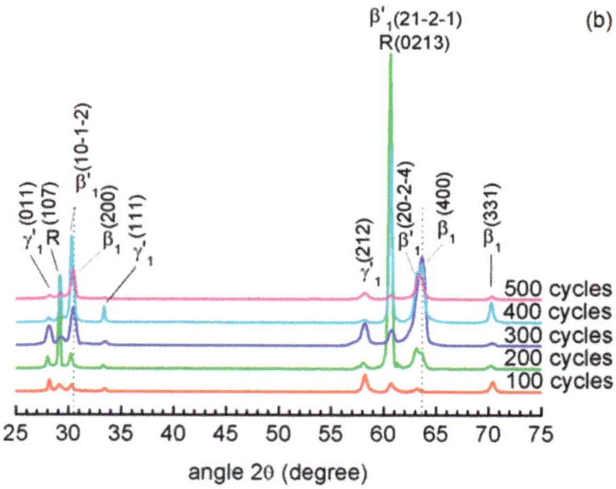

Figure 11. Diffractograms of Cu-13.7% Al-4.2% Ni alloy after TCTs under loading of 0.53 MPa finished with the (**a**) 1/2 cooling cycle and (**b**) 1/2 heating cycle.

The 300 cycle TCTs promoted greater stability of the martensite γ'_1 identified by its $(011)\gamma'_1$ and $(212)\gamma'_1$ peaks. The coherent plane $\beta'_1|(0213)$R peak between the martensitic phases β'_1 and R, and $(331)\beta_1$ peak of the high-temperature phase β_1 were detected with moderate intensity. The $(10\text{-}1\text{-}2)\beta'_1$, $(20\text{-}2\text{-}4)\beta'_1$, and (107)R peaks were revealed with low intensity. After 400 cycles, the amount of martensitic β'_1 phase increased, exhibiting intensive peaks $(10\text{-}1\text{-}2)\beta'_1$ and $(20\text{-}2\text{-}4)\beta'_1$. The coherent plane $\beta'_1|(0213)$R peak between the martensitic β'_1 and R-phases, $(331)\beta_1$ peak of the high-temperature β_1 phase, and the (107)R peak of R-phase were detected with moderate intensity. Martensite γ'_1 was detected in a small amount.

After 500 TCTs, the XRD pattern exhibited the prominent participation of the martensitic β'_1 phase by its $(10\text{-}1\text{-}2)\beta'_1$ and $(20\text{-}2\text{-}4)\beta'_1$ peaks while decreasing all other peaks related to γ'_1, R, and $\beta'_1|$R-phases.

3.3.6. XRD Analysis after TCT under Loading of 0.53 MPa Ending with 1/2 Heating Cycle

The Cu-13.7% Al-4.2% Ni alloy diffractograms after TCTs finished with the 1/2 heating cycle are shown in Figure 11b. After 100 cycles, major participation by the martensitic γ'_1 phase, identified by intensive $(011)\gamma'_1$, $(111)\gamma'_1$, $(212)\gamma'_1$ peaks was observed. The R, β'_1, $\beta'_1|$R, and β_1 phases were detected by moderate intensity (107)R, $(10\text{-}1\text{-}2)\beta'_1$, $(20\text{-}2\text{-}4)\beta'_1$, $\beta'_1|(0213)$R, and $(331)\beta_1$ peaks. With the accumulation of 200 TCTs, the amount of γ'_1 martensite decreased while the participation of the coherent plane between the martensitic β'_1 and R-phases, $\beta'_1|(0213)$R, increased. Associated with the R-phase, the (107)R peak was detected with moderate intensity. There was an insignificant amount of the other aforementioned phases.

After 200 cycles, the martensite β'_1 stabilized, its peaks$(10\text{-}1\text{-}2)\beta'_1$ and $(20\text{-}2\text{-}4)\beta'_1$ shifted slightly, thus developing a strong coherence with the high-temperature phase β_1, represented by $(200)\beta_1$ and $(400)\beta_1$ peaks. After 400 TCTs, the R, γ'_1, and β_1 phases were detected in smaller amounts.

3.4. Optical Microscopy after TCT under Loading

The structure of Cu-13.7% Al-4.2% Ni alloy after TCTs under the load of 0.11 MPa is shown in Figures 12 and 13. In the initial state, the alloy demonstrated a surface structure composed of martensitic blocks: a central block and four peripheral blocks. No significant changes were detected by optical microscopy at the end of treatment and 1/2 cooling cycle (Figure 12a,c,e and Figure 13a,c) or 1/2 heating cycle (Figure 12b,d,f and Figure 13b,d).

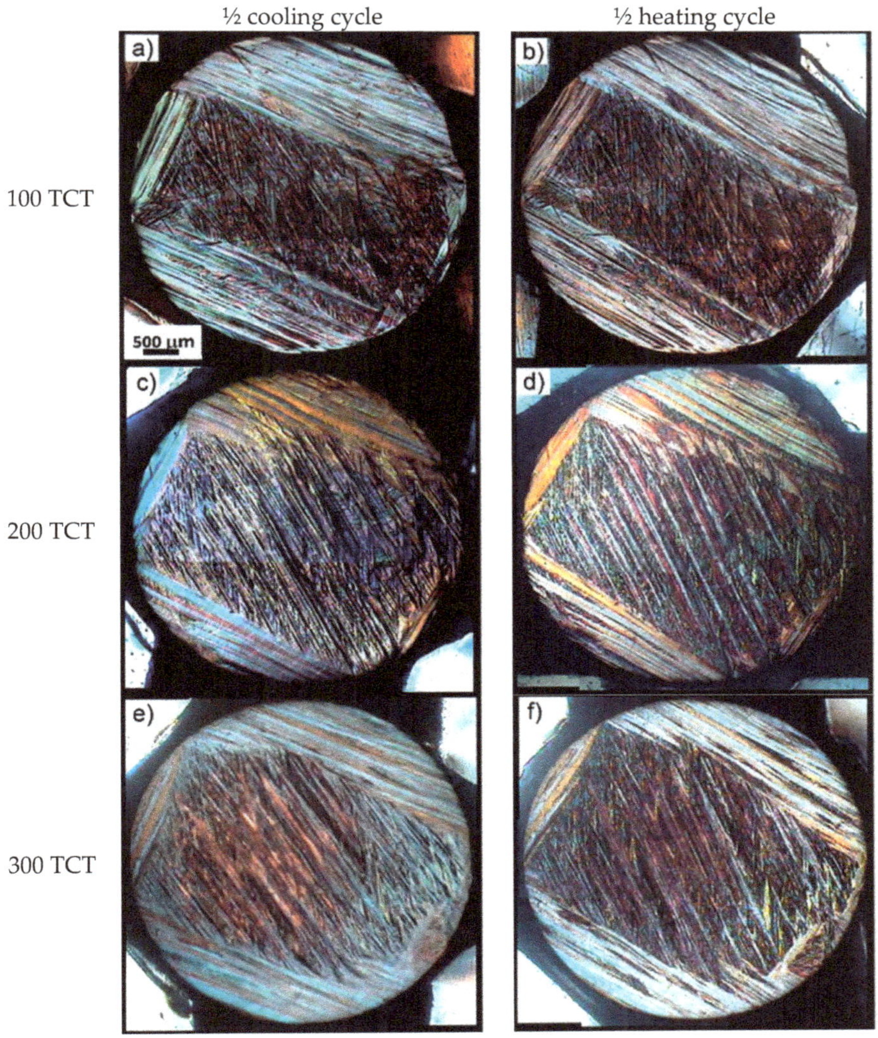

Figure 12. Microstructure of Cu-13.7% Al-4.2% Ni alloy after (**a**,**b**) 100; (**c**,**d**) 200; and (**e**,**f**) 300 TCTs under the loading of 0.11 MPa, finished with the (**a**,**c**,**e**) 1/2 cooling cycle and (**b**,**d**,**f**) 1/2 heating cycle.

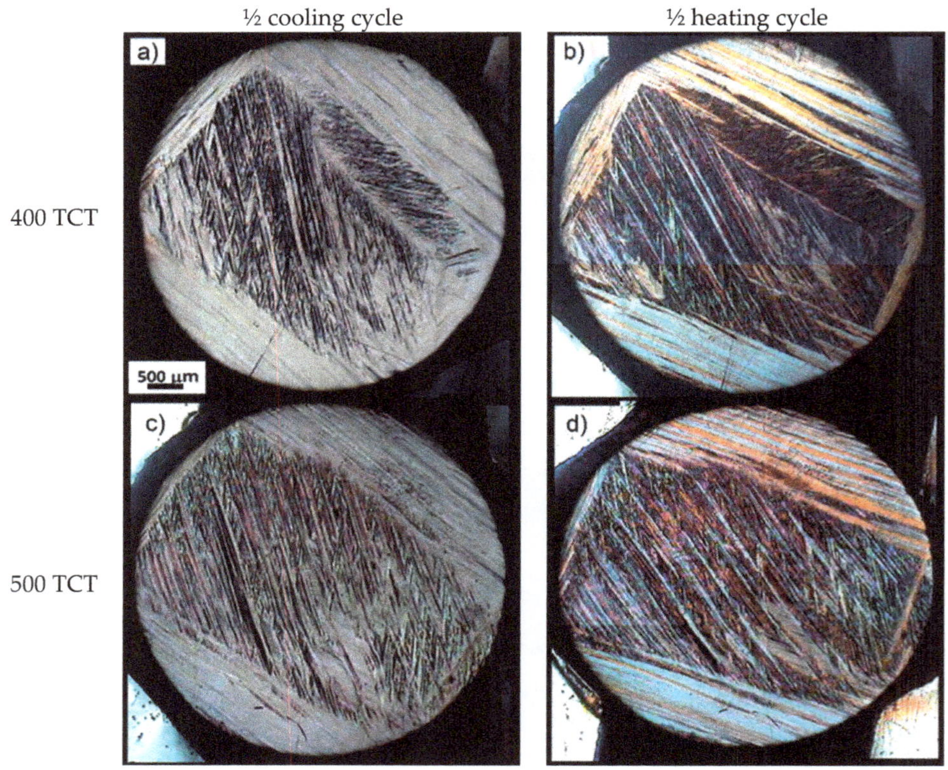

Figure 13. Microstructure of Cu-13.7% Al-4.2% Ni alloy after (**a**,**b**) 400 and (**c**,**d**) 500 TCTs under the loading of 0.11 MPa, finished with the (**a**,**c**) 1/2 cooling cycle and (**b**,**d**) 1/2 heating cycle.

After TCTs under the load of 0.11 MPa, two martensitic blocks of the peripheral region were eliminated (Figure 12a–f). With the accumulation of 400 and 500 cycles (Figure 13a–d), the martensitic lamellae were not as distinct as before, but it was possible to examine microscopic characteristics of fine needles of this β'_1 martensitic phase as estimated by XRD (Figure 9) [1,3].

The microstructure of Cu-13.7% Al-4.2% Ni alloy after TCTs under the load of 0.26 MPa is shown in Figures 14 and 15. The accumulation of structural imperfections during 100 and 200 cycles promoted the development of deep sliding lines, superimposed on the specific martensitic relief (Figure 14a,c). These lines originated from the multiple RMTs induced by TCTs [3,8,18–22]. On heating, the superficial lines became microscopically visible due to the relief of tension (Figure 14b,d). In the peripheral part a spear-shaped needle feature, martensite γ'_1 (highlighted in Figure 14b) was presented prominently in the XRD pattern of Figure 10b. With the accumulation of 300 thermal cycles, the reorientation of the martensite lamellae could be noticed (Figure 14e,f). The microstructure of the alloy after TCTs of 400 and 500 cycles was practically stable, with the completely oriented martensitic lamellae in the central region (Figure 15a–d).

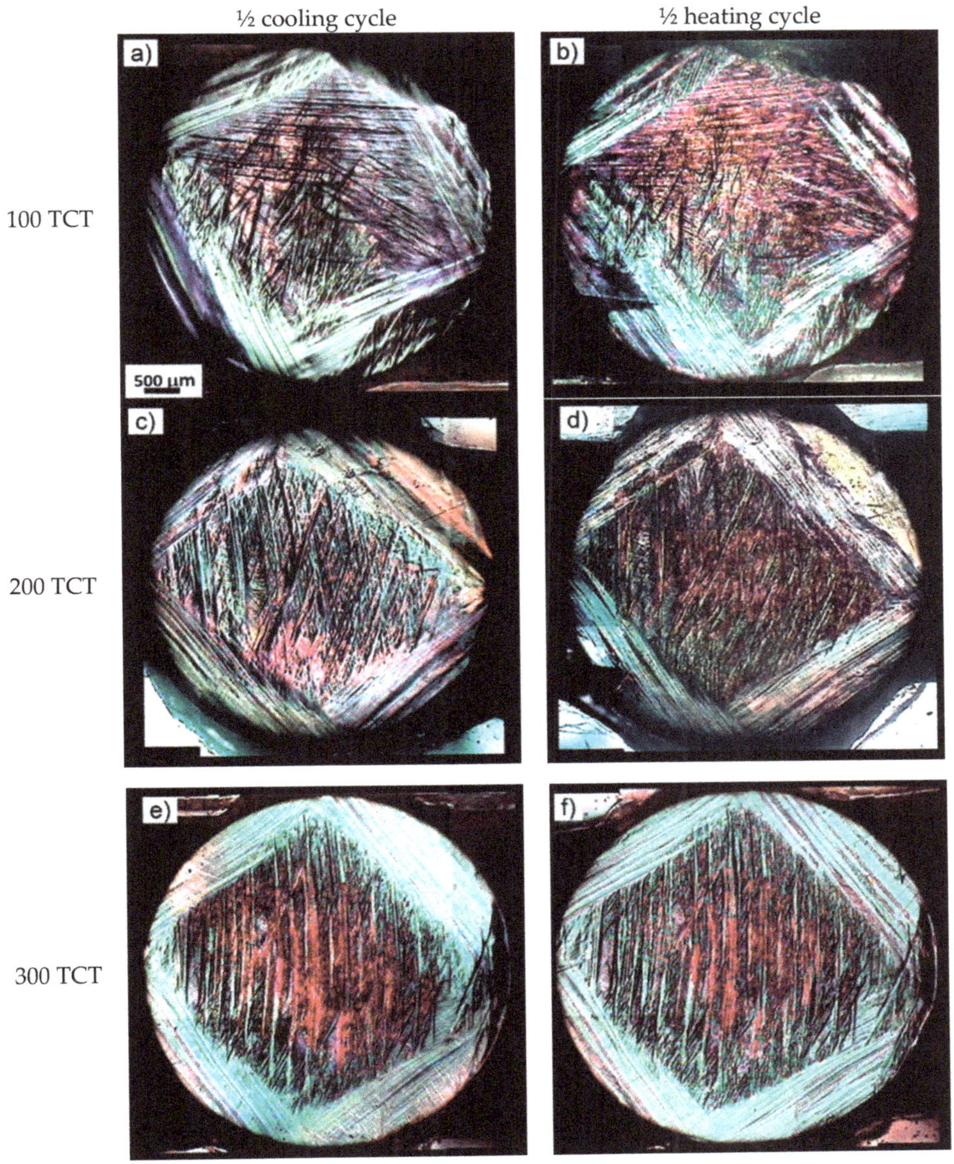

Figure 14. Microstructure of Cu-13.7% Al-4.2% Ni alloy after (**a**,**b**) 100; (**c**,**d**) 200; and (**e**,**f**) 300 TCTs under the loading of 0.26 MPa, finished with the (**a**,**c**,**e**) 1/2 cooling cycle and (**b**,**d**,**f**) 1/2 heating cycle.

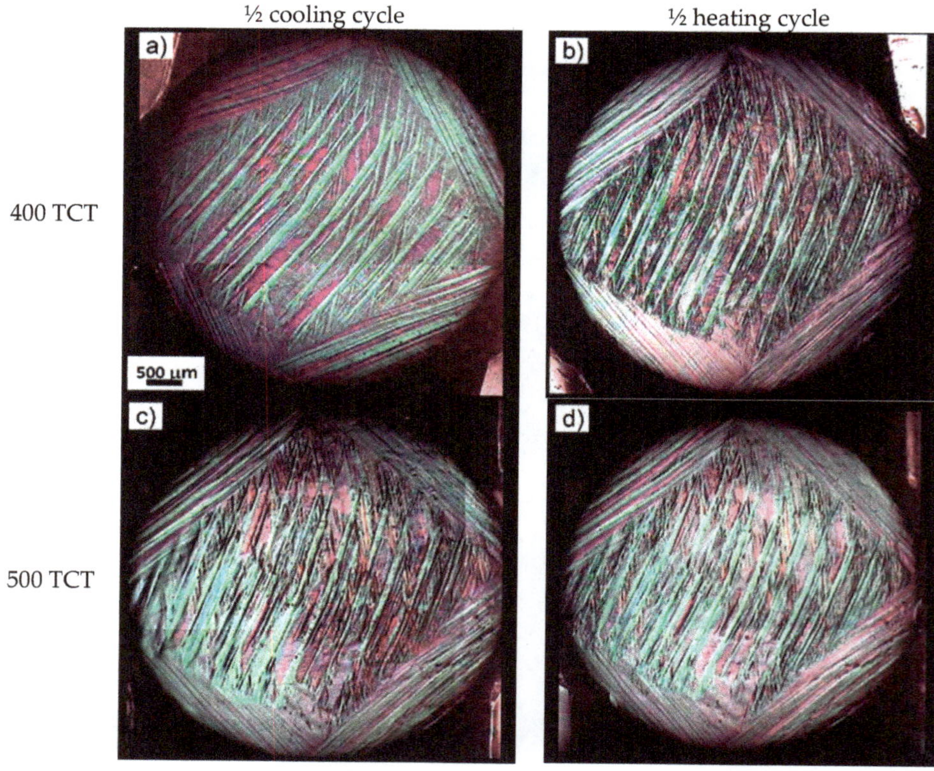

Figure 15. Microstructure of Cu-13.7% Al-4.2% Ni alloy after (**a**,**b**) 400 and (**c**,**d**) 500 TCTs under the loading of 0.26 MPa, finished with the (**a**,**c**) 1/2 cooling cycle and (**b**,**d**) 1/2 heating cycle.

The morphological aspect of Cu-13.7% Al-4.2% Ni alloy after TCT under the load of 0.53 MPa is shown in Figures 16 and 17. With the accumulation of thermal cycles, the martensitic relief was less pronounced, probably due to the coherency between the martensitic β'_1 and R-phases as noted in the diffractograms of Figure 11. With the accumulation of 300, 400, and 500 cycles, the martensitic relief changed; thus, differently oriented lamellae could be distinguished within the initial martensitic blocks (Figure 16e,f and Figure 17a,d). After TCTs, the individual blocks of the martensitic lamellae with different orientations were observed throughout the sample's cross-sectional area. It is noted that the sliding lines caused by the RMT overlapped in certain regions, where the accumulation of structural defects probably occurs during TCT tests. Nevertheless, it can be said that the structural orientation was the result of the training under the applied load.

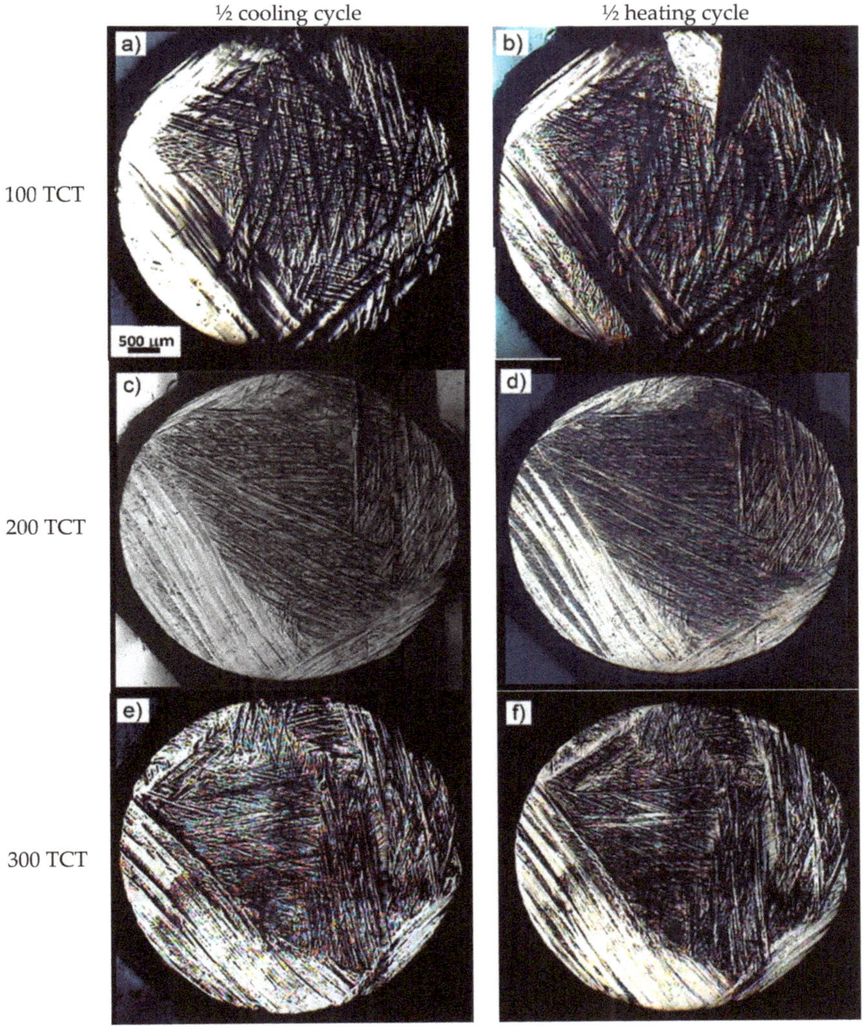

Figure 16. Microstructure of Cu-13.7% Al-4.2% Ni alloy after (**a**,**b**) 100; (**c**,**d**) 200; and (**e**,**f**) 300 TCTs under the loading of 0.53 MPa, finished with the (**a**,**c**,**e**) 1/2 cooling cycle and (**b**,**d**,**f**) 1/2 heating cycle.

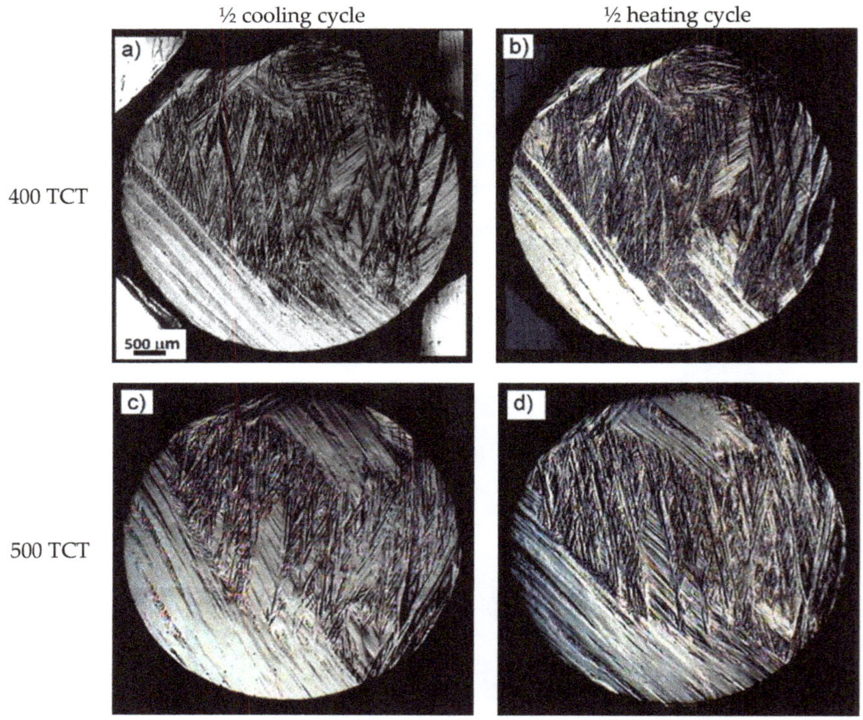

Figure 17. Microstructure of Cu-13.7% Al-4.2% Ni alloy after (**a**,**b**) 400 and (**c**,**d**) 500 TCTs under the loading of 0.53 MPa, finished with the (**a**,**c**) 1/2 cooling cycle and (**b**,**d**) 1/2 heating cycle.

3.5. Vickers Microhardness after TCT under Loads

The average values of Vickers microhardness as a function of the number of thermal cycles under the loads of 0.11, 0.26, and 0.53 MPa applied to the monocrystalline Cu-13.7% Al-4.2% Ni alloy are shown in Figure 18. Determined as 303 ± 7.8 kgf/mm^2 in the initial state, microhardness tended to decrease after TCTs under the loads of 0.11 and 0.53 MPa, going down to 241.8 ± 8.1 kgf/mm^2 and 268.7 ± 6 kgf/mm^2, respectively, after 500 TCTs. However, after TCTs under the load of 0.26 MPa, microhardness displayed a stability plateau up to 300 thermal cycles. With the higher number of cycles, the microhardness increased from 295.3 ± 7.9 kgf/mm^2, after 300 cycles, to 341.9 ± 14.5 kgf /mm^2, after 500 thermal cycles, which could be typical for the more strained condition of any alloy with a stable structure due to the accumulation of imperfections that can hinder the development of the transformation. But, the apparent increase in microhardness can be associated with the recovery during unloading of the reversible strain accumulated by the induced martensitic transformation [1–3].

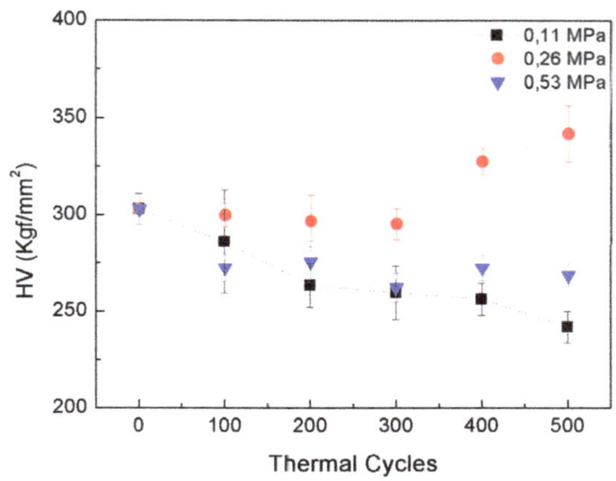

Figure 18. Vickers microhardness as function of the number of thermal cycles under loadings of 0.11, 0.26, and 0.53 MPa applied to the monocrystalline Cu-13.7% Al-4.2% Ni alloy.

4. Discussion

The investigated alloy demonstrated complex phase transformation involving coherent phases sensitive to experimental conditions. One of the key proofs of the coherence among the phases in the alloy structure was the absence of several DSC transformation peaks, contrary to that observed in similar alloys [3,24–29]. A single RMT peak indicated the sequence of transformation $\beta_1 \leftrightarrow R \leftrightarrow \beta'_1 + \gamma'_1$ in the structure of monocrystalline Cu-13.7% Al-4,2% Ni alloy.

It should be noted that after TCTs finishing with the $\frac{1}{2}$ heating cycle (AT → 90 °C → AT without loading), the XRD patterns of the alloys (Figures 9b, 10b and 11b) were in good agreement with the direct RMT critical temperatures (Figures 5a, 6a and 7a). Figure 5a shows that the temperature of the direct intensive RMT M_{of}, determined after the first thermal cycle of the alloy, was above the TA, estimated by XRD analysis (20 ± 5 °C); it is assumed that this is within the martensite γ'_1 field (below M_{of}) as observed in the diffractogram of Figure 9b. In general, when the critical temperature M_{of} decreased approaching AT, a greater participation of the martensite β'_1 and less participation of the others phases (γ'_1, $\beta'_1|R$ and R) was observed in the diffractograms of Figures 9b, 10b and 11b.

It is noted that during all TCTs (Figures 5b, 6b and 7b), the XRD procedure for temperature (20 ± 5 °C) went beyond the reverse RMT temperatures even with the accumulation of TCTs, and the critical temperature A_f was very far from TA. So, the participation of the high-temperature phase β_1 (mainly in the diffractograms after TCTs under the load of 0.26 MPa (Figure 10) was assumed to be a load response that promoted reverse RMT: $\gamma'_1 + \beta'_1 \rightarrow R \rightarrow \beta_1$, only partially. These results are in agreement with our previous work [18–22], where the same TCTs of a Cu–Al–Ni alloy with comparable composition were performed; however, in TCTs without the applied load, it was not noticed.

During thermal cycling under load, two thermodynamic factors had a complex effect on the implementation of reversible martensitic transformations. Firstly, the load induced reversible MTs of the M → A_{def} type, when heated from the martensitic phase temperature, or A → M_{def}, when cooled from the austenitic phase temperature, thus orienting the structure in a certain way [1–3,32]. The fact of the possible reversible martensitic transformation of M- into A-phase under load, as predicted by Vasilevsky in 1971 [32], was experimentally proven in the studies on TiNi- [33–37] and Cu-Al-Ni [38]-based alloys with RMT.

Secondly, the load induced deformation that can be reversible (SME) or irreversible (inelastic deformation) in the alloys with RMT [2,3]. Reversible deformation depends on the load and it increases to a certain value (4–8%) initially but later on decreases. In this case, the temperature ranges for the realization of RMT and SMA expanded drastically as would be studied for a TiNi alloy [33–37].

According to the Clausius–Clapeyron equation, the applied external stress resulted in an increase in the critical temperatures of RMT [3,39]. On the other hand, the directed action of the load promoted some orientation of the structure and facilitated the transformation of martensitic lamellae according to certain variants [3,5].

As early as the 1980s in the works of Antipov et al. [40], the behavior of TiNi alloys with SME subjected to TCTs under different loads of flexural strain was analyzed. It was discovered that on heating during TCT, the return angle (determines the degree of shape restoration) decreased intensively in the first cycles and stabilized an increase in the number of thermal cycles, revealing the stabilization of the structure accompanied by a martensite reorientation and hardening. It was found that under optimized stresses, the return angle stabilization and stability of the structure develops with fewer cycles, while under low and/or excessive applied stresses, angle stabilization is delayed or even noticeable.

Comparison of the results of Antipov et al. [40] with the results of this work for the Cu–Al–Ni SMAs reveals a very similar tendency. With TCTs under the lowest applied load (0.11 MPa), structural stability with reorientation and hardening was not achieved even after TCT of 500 cycles, although critical temperatures and hysteresis showed fewer changes during cyclic tests as observed in Figures 5–8.

During TCTs under a higher applied load (0.26 MPa), RMT-induced processes of reorientation, hardening, and stabilization of the structure occurred more intensively, resulting in critical intervals reduction, microhardness increasing, and stabilization, from 300 thermal cycles. Such a conclusion was based on microhardness behavior, critical temperatures, hysteresis, and structural changes as observed in Figures 6, 8 and 18.

With the highest applied load (0.53 MPa), the reorientation and hardening processes could be expected to be faster; however, the excessive load acts both to facilitate the reorientation and to make the shape restoration difficult. As a result, the entire volume of alloy does not participate in the RMT in the limited temperature range adopted in the tests. In this process, the accumulation was more significant which slows down all the RMT reactions, increasing the critical intervals and revealing the intermediate structural states at the ambient temperature (TA) as observed in Figures 7 and 8.

Nevertheless, the changes in the main RMT parameters of monocrystalline Cu-13.7% Al-4.2% Ni alloy after TCT under all applied loads (after "training" under loads) were not very significant, especially after complete RMT cycles which indicates good resistance of the alloy investigated to irreversible structure changes.

5. Conclusions

1. The monocrystalline Cu-13.7% Al-4.2% Ni alloy represents complex mixed transformation, revealing the RMT sequence $\beta_1 \leftrightarrow R \leftrightarrow \beta'_1 + \gamma'_1$, where the present phases are coherent with each other and are very sensitive to structural changes.
2. Alterations in the alloy structure by finishing the 1/2 cycle heating treatment (TA → 90 °C → TA) show good consistency with the critical temperatures of direct RMT. The martensite γ'_1 field lies below the end of the intensive direct RMT (below M_{of}). Above this temperature, a greater participation of martensite β'_1 is observed, coherent to phase R.
3. The participation of the high-temperature phase β_1 in the alloy structure, mainly after TCTs under the load of 0.26 MPa, appears as a response of the load applied during the treatment which promotes partially the reverse RMT $\gamma'_1 + \beta'_1 \rightarrow R \rightarrow \beta_1$.
4. During TCTs performed under optimized load (0.26 MPa), where the RMT range was not changed, the reorientation, hardening, and stabilization of the structure during RMT occur more intensively, resulting in a reduction of critical intervals, increasing and stabilizing of the microhardness, from 300 thermal cycles.
5. The changes in the main RMT parameters of monocrystalline Cu-13.7% Al-4.2% Ni alloy during TCT under load are not very significant, especially after complete RMT cycles, which indicate good resistance of the alloy investigated to irreversible changes, making its practical use feasible.

Author Contributions: L.A.M.: conceptualization, project administration, creation of the special device for heat cycling, analyses of all results, original manuscript preparation; E.C.P.: investigation, the heat cycling tests performing, structure and DSC analyses, original figures preparation, discussion; S.A.P.: support and manufacture of the investigated alloy; C.Y.S.: DSC performing and DSC analysis, N.A.P.: discussion, translation into English and article edition, provision of referenced papers, including links and DOI. All authors have read and agreed to the published version of the manuscript.

Funding: The researchers are grateful for the research support granted by FAPERJ, CAPES, and CNPq; N.A.P. thanks the support of state assignment No. 075-00947-20-00.

Acknowledgments: The researchers are grateful for the research support granted by Luciana Lezira Pereira de Almeida technique of LAMAV/CCT/UENF for the performance of X-ray diffraction analysis.

Conflicts of Interest: The authors declare no conflict of interest. The funders had no role in the design of the study; in the collection, analyses, or interpretation of data; in the writing of the manuscript, or in the decision to publish the results.

References

1. Kurdyumov, G.V.; Khandros, L.G. On the thermoelastic equilibrium on martensitic transformations. *Dokl. Akad. Nauk SSSR* **1949**, *66*, 211–214. Available online: http://archive.ujp.bitp.kiev.ua/files/journals/53/si/53SI20p.pdf (accessed on 25 January 2020).
2. Perkins, J. Shape memory effects in alloys. In Proceedings of the International Symposium on Shape Memory Effects and Applications, Toronto, ON, Canada, 19–20 May 1975; Available online: https://www.springer.com/gp/book/9781468422139 (accessed on 25 January 2020).
3. Otsuka, K.; Wayman, C.M. (Eds.) *Shape Memory Materials*; Cambridge University Press: Cambridge, UK, 1999; Available online: https://ru.scribd.com/document/368059783/K-Otsuka-C-M-Wayman-Shape-Memory-Materials-Cambridge-University-Press-1998 (accessed on 25 January 2020).
4. Duerig, T.W.; Pelton, A.; Stöckel, D. An overview of nitinol medical applications. *Mater. Sci. Eng. A* **1999**, *273*, 149–160. [CrossRef]
5. Otsuka, K.; Ren, X. Physical metallurgy of Ti–Ni-based shape memory alloys. *Prog. Mater. Sci.* **2005**, *50*, 511–678. [CrossRef]
6. SMST-2017: Nitinol and Beyond. In Proceedings of the International Conference on Shape Memory and Superelastic Technologies (SMST 2017), San Diego, CA, USA, 15–19 May 2017; Volume 1, p. 230. Available online: https://www.asminternational.org/web/smst/events/proceedings/-/journal_content/56/10192/37425484/PUBLICATION (accessed on 25 January 2020).
7. Font, J.; Cesari, E.; Muntasell, J.; Pons, J. Thermomechanical cycling in Cu–Al–Ni-based melt-spun shape-memory ribbons. *Mater. Sci. Eng. A* **2003**, *354*, 207–211. [CrossRef]
8. Tadaki, T. Cu-based shape memory alloys. In *Shape Memory Materials*; Otsuka, K., Wayman, C.M., Eds.; Cambridge University Press: Cambridge, UK, 1999; pp. 97–116.
9. Martynov, V.; Bokate, M.D.; Johnson, A.D.; Grey, G.R. Superelastic and shape memory single Crystal Cu-Al-Ni: Fabrication and Applications. In Proceedings of the International Conference on Shape Memory and Superelastic Technologies, Pacific Grove, CA, USA, 7–11 May 2006; pp. 427–439. Available online: http://docplayer.net/50163627-Smst-shape-memory-and-superelastic-technologies-proceedings-of-the-international-conference-on.html (accessed on 25 January 2020).
10. Pulnev, S.; Nikolaev, V.; Priadko, A.; Rogov, A.; Vahhi, I. Actuators and drivers based on CuAlNi shape memory single crystals. *J. Mater. Eng. Perform.* **2011**, *20*, 497–499. [CrossRef]
11. Morin, M.; Trivero, F. Influence of thermal cycling on the reversible martensitic transformation in a Cu-Al-Ni shape memory alloy. *Mater. Sci. Eng. A* **1995**, *196*, 177–181. [CrossRef]
12. Silva, R.J.C.; Matlakhova, L.A.; Pereira, E.C.; Matlakhov, A.N.; Monteiro, S.N.; Rodríguez, R.J.S. Thermal cycling treatment and structural changes in Cu-Al-Ni monocrystalline alloys. *Mater. Sci. Forum Trans. Tech. Publ.* **2006**, *514*, 692–696. [CrossRef]
13. Matlakhova, L.A.; Motta, A.C.; Matlakhov, A.N.; Pereira, E.C.; Toledo, R.; Monteiro, S.N. Estrutura de ligas monocristalinas Cu-Al-Ni após ciclagem térmica. *Technol. Metall. Mater. Min.* **2009**, *6*, 80–85.
14. Hund, F.; Fricke, R. Der Kristallbau von α-BiF3. *Z. Anorg. Chem.* **1949**, *258*, 198–204. [CrossRef]
15. Vol, A.E. *The Structure and Properties of Binary Systems*; Fizmatgiz: Moscow, Russia, 1959; Volume 1, p. 755.
16. Bown, M.G. The structure of rhombohedral T (NiCuAl). *Acta Crystallogr.* **1956**, *9*, 70–74. [CrossRef]

17. Shishkovsky, I.; Yadroitsev, I.; Morozov, Y. Laser-assisted synthesis in Cu-Al-Ni system and some of its properties. *J. Alloys Compd.* **2016**, *658*, 875–879. [CrossRef]
18. Matlakhova, L.A.; Pereira, E.C.; Matlakhov, A.N.; Monteiro, S.N. Structure and properties of a monocrystalline Cu-Al-Ni alloy submitted to thermal cycling under load. In *Shape Memory Alloys: Manufacture, Properties and Applications*; Nova Publishers: Hauppauge, NY, USA, 2010; ISBN 978-1-60741-789-7. Available online: http://www.novapublishers.org/catalog/product_info.php?products_id=20903 (accessed on 25 January 2020).
19. Pereira, E.C.; Matlakhova, L.A.; Matlakhov, A.N.; Monteiro, S.N.; Rodriguez, R.S. Effect of applied load on the characteristics of reversible martensitic transformation during thermal cycling treatment of a monocrystalline Cu-13.5Al-4Ni alloy. *Mater. Sci. Res. J.* **2010**, *4*, 15–36. Available online: https://www.researchgate.net/publication/286317945_Effect_of_applied_load_on_the_characteristics_of_reversible_martensitic_transformation_during_thermal_cycling_treatment_of_a_monocrystalline_Cu-13.5Al-4Ni_alloy (accessed on 25 January 2020).
20. Matlakhova, L.A.; Pereira, E.C.; Matlakhov, A.N.; Monteiro, S.N.; Rodriguez, R.S. Stress assisted thermal cycling treatment of a monocrystalline Cu-Al-Ni alloy. *Mater. Sci. Res. J.* **2011**, *5*, 31–50.
21. Pereira, E.C.; Matlakhova, L.A.; Matlakhov, A.N.; Shigue, C.Y.; Monteiro, S.N. Physical and Structural Characterization of a Monocrystalline Cu-13.7Al-4.2Ni Alloy Subjected to Thermal Cycling Treatments. *Metall. Mater. Trans. A* **2014**, *45*, 1866–1875. [CrossRef]
22. Pereira, E.C.; Matlakhova, L.A.; Matlakhov, A.N.; Araújo, C.J.; Shigue, C.Y.; Monteiro, S.N. Reversible martensite transformations in thermal cycled polycrystalline Cu-13.7% Al-4.0% Ni alloy. *J. Alloys Compd.* **2016**, *688*, 436–446. [CrossRef]
23. Pul'nev, S.A.; Nikolaev, V.I.; Malygin, G.A.; Kuz'min, S.L.; Shpeizman, V.V.; Nikanorov, S.P. Generation and relaxation of reactive stresses in Cu-Al-Ni shape-memory alloy. *Tech. Phys.* **2006**, *51*, 1004–1007. [CrossRef]
24. Recarte, V.; Pérez-Sáez, R.B.; Bocanegra, E.H.; Nó, M.L.; San Juan, J. Dependence of the martensitic transformation characteristics on concentration in Cu–Al–Ni shape memory alloys. *Mater. Sci. Eng. A* **1999**, *273*, 380–384. [CrossRef]
25. Recarte, V.; Pérez-Sáez, R.B.; Nó, M.L.; San Juan, J. Evolution of martensitic transformation in Cu–Al–Ni shape memory alloys during low-temperature aging. *J. Mater. Res.* **1999**, *14*, 2806–2813. [CrossRef]
26. Recarte, V.; Pérez-Sáez, R.B.; Nó, M.L.; San Juan, J. Ordering kinetics in Cu–Al–Ni shape memory alloys. *J. Appl. Phys.* **1999**, *86*, 5467–5473. [CrossRef]
27. Recarte, V.; Pérez-Sáez, R.B.; San Juan, J.; Bocanegra, E.H.; Nó, M.L. Influence of Al and Ni concentration on the martensitic transformation in Cu-Al-Ni shape-memory alloys. *Metall. Mater. Trans. A* **2002**, *33*, 2581–2591. [CrossRef]
28. Ibarra, A.; Caillard, D.; San Juan, J.; Nó, M.L. Martensite nucleation on dislocations in Cu–Al–Ni shape memory alloys. *Appl. Phys. Lett.* **2007**, *90*, 101–107. [CrossRef]
29. Ibarra, A.; San Juan, J.; Bocanegra, E.H.; Nó, M.L. Evolution of microstructure and thermomechanical properties during superelastic compression cycling in Cu–Al–Ni single crystals. *Acta Mater.* **2007**, *55*, 4789–4798. [CrossRef]
30. Nishiyama, Z. *Martensitic Transformations*; Fine, M.E., Meshii, M., Wayman, C.M., Eds.; Academic Press: San Francisco, NY, USA, 1978; p. 480.
31. Swann, P.R.; Warlimont, H. The electron-metallography and crystallography of copper-aluminum martensites. *Acta Metall.* **1963**, *11*, 511–527. [CrossRef]
32. Wasilewski, R.J. The effects of applied stress on the martensitic transformation in TiNi. *Metall. Mater. Trans.* **1971**, *2*, 2973–2981. [CrossRef]
33. Kovneristy, Y.K.; Fedotov, S.G.; Matlakhova, L.A. Effect of deformation on phase transformations and modulus of elongation in titanium nickelide—Based alloy. In Proceedings of the 5th International Conference on Titanium, Munich, Germany, 10–14 September 1984; Volume 3, pp. 1673–1681.
34. Kovneristy, Y.K.; Fedotov, S.G.; Matlakhova, L.A.; Oleynikova, S.V. Shape memory and shape reversibility effects in a TiNi alloy as a function of deformation. *Fiz. Met. Metallogr. (USSR)* **1986**, *62*, 124–129.
35. Kovneristy, Y.K.; Fedotov, S.G.; Matlakhova, L.A. The influence of plastic deformation on the structure, shape memory effect and other properties of TiNi alloy. In Proceedings of the International Symposium CN Shape Memory Alloys, Guilin, China, 6–9 September 1986; pp. 175–180.
36. Fedotov, S.G.; Kovneristy, Y.K.; Matlakhova, L.A.; Zhebynieva, N.F. Structural changes in alloy TiNi with the shape memory effect during deformation. *Phys. Met. Metall.* **1988**, *65*, 564–569.

37. Matlakhova, L.A.; Belousov, O.K.; Palii, N.A. Elastic Modulus and structure changes in TiNi during SME and RSME. Available online: https://www.lap-publishing.com/catalog/details//store/gb/book/978-613-9-85937-5/elastic-modulus-and-structure-changes-in-tini-during-sme-and-rsme?locale=ru (accessed on 25 January 2020).
38. Matlakhova, L.A.; Motta, A.C.; Matlakhov, A.N.; Kolmakov, A.G.; Sevostianov, M.A. Módulo de elasticidade da liga monocristalina Cu-Al-Ni deformada por compressão. In Proceedings of the 18th Brazilian Congress of Engineering and Materials Science—CBECiMat, Porto de Galinhas, Brazil, 24–28 November 2008; pp. 6463–6474. Available online: https://www.scribd.com/document/50128870/18Cbecimat (accessed on 25 January 2020).
39. Pérez-Landazábal, J.I.; Recarte, V.; Pérez-Sáez, R.B.; Nó, M.L.; Campo, J.; San Juan, J. Determination of the next-nearest neighbor order in β phase in Cu-Al-Ni shape memory alloys. *Appl. Phys. Lett.* **2002**, *81*, 1794–1796. [CrossRef]
40. Antipov, A.I.; Matlakhova, L.A.; Tolmachev, A.N.; Oleinikova, S.V.; Zebyneva, N.F. Effect of Thermomechanical Treatment on the Shape Memory Effect in TN-1 Alloys. *Fiz. Khim. Obrab. Mater.* **1981**, *6*, 67–70.

metals

MDPI

Article

Study on Phase Transformation in Hot Stamping Process of USIBOR® 1500 High-Strength Steel

Pengyun Zhang [1], Le Zhu [2], Chenyang Xi [3] and Junting Luo [3,*]

1 School of Information Technology, Hebei University of Economics and Business, Shijiazhuang 050061, China; zhangpengyun1982@126.com
2 Industrial Technology Centre, Chengde Petroleum College, Chengde 067000, China; 15990095530@163.com
3 State key laboratory of Metastable Materials Science and Technology, Yanshan University, Qinhuangdao 066004, China; xichenxu123@163 com
* Correspondence: luojunting@ysu.edu.cn; Tel.: −86-335-805-2253

Received: 17 September 2019; Accepted: 17 October 2019; Published: 19 October 2019

Abstract: Based on the Kirkaldy-Venugopalan model, a theoretical model for the phase transformation of USIBOR® 1500 high strength steel was established, and a graph of the phase transformation kinetics of ferrite, pearlite, and bainite were plotted using the software MATLAB. Meanwhile, with the use of the software DYNAFORM, the thermal stamping process of an automobile collision avoidance beam was simulated. The phase transformation law of USIBOR® 1500 high-strength steel during hot stamping was studied through a simulation of the phase transformation during the pressure holding quenching process. In combination with the continuous cooling transformation (CCT) curve, the cooling rate of quenching must be greater than 27 °C/s to ensure maximum martensite content in the final parts, and the final martensite content increases as the initial temperature of the sheet rises.

Keywords: high-strength steel; hot stamping; martensitic transformation; finite element analysis

1. Introduction

Advanced high-strength steel (AHSS) usually refers to automotive steel with a strength of more than 500 MPa and good plasticity [1], such as dual-phase steel, transformation induced plasticity steel, martensitic steel, hot-formed steel, and twinning induced plasticity steel. ASHH has excellent mechanical properties in normalized or normalized and tempered conditions. The application of AHSS is an effective way to achieve lightweight and high safety of automobiles. It has been widely used in the automotive industry, mainly for automotive structural parts, safety parts, and reinforcements, such as A/B/C columns, front and rear bumpers, door anti-collision beams, beams, stringers, seat slides, etc. [2]. The tensile strength of USIBOR® 1500 high-strength steel after rolling annealing is approximately 400–600 MPa. The uniform austenite structure is obtained by heating in a furnace to about 900 °C and holding for several minutes. Then, the sheet is formed in a mold. After forming and quenching in the mold, the complete martensite structure can be obtained, and the strength of such martensitic microstructure steel can exceed 1500 MPa at room temperature [3,4]. The hot stamping process of high-strength steel sheets is a thermal-force-phase transformation coupling process. The change of temperature affects the mechanical properties of the material, and the deformation and temperature affect the phase change simultaneously, thereby ultimately determining the quality of the formed parts. Quenching at different cooling rates can induce different forms of phase transformation, such as ferrite, retained austenite, bainite, and martensitic transformation. Therefore, the mechanism of the phase transformation of martensite during hot stamping must be studied for achieving ultra-high-strength materials.

Finite element analysis is a common method of studying the phase change process and mechanism [5]. Wang et al. [6] studied the effect of austenite's holding time on phase transformation by finite element analysis. Simulation results showed that the final martensite volume fraction changed with an increase of the holding time in the austenite temperature range of 800–900 °C. On the basis of phase change martensite theory, Zhu et al. [7] established a three-dimensional model for the study of hot stamping simulation. The simulation results showed that the temperature distribution between the mold gap and the mold considerably influenced the microstructure of the high-strength steel at the end of the process. Tamas et al. [8] predicted microstructure transformation and mechanical properties by establishing a dynamic model of the quenching process of low-alloy steels considering the effect of austenite grain growth rate. Bok et al. [9] used finite element software to simulate the microstructure transformation in hot stamping according to three phase transformation models and then predicted sheet hardness under the different models.

For diffusion phase transformation, the Johnson-Mehl-Avrami-Kolmogorov (JMAK) equation accurately expresses the relationship between the amount of microstructure transformation and time and temperature, but it does not describe well the degree and law of the influence of various influencing factors on microstructure transformation. The Kirkaldy-Venugopalan model is a good descriptor of the effects of major factors on microstructure transformation, but the accuracy is worse than the JMAK equation. In order to study the phase transformation of USIBOR® 1500 high strength steel, the Kirkaldy-Venugopalan model [10,11] is selected and modified to establish a theoretical model for the phase transformation of USIBOR® 1500 high-strength steel, which is closer to the actual phase transformation kinetic curve.

On the basis of establishing a theoretical model of the phase change of USIBOR® 1500 high-strength steel, this work regards the thermal stamping process of an automobile collision-proof beam as an example. Through the simulation of the phase change process in pressure retaining and quenching, the phase change law in the hot stamping process of USIBOR® 1500 high-strength steel is studied.

2. Material and Methods

USIBOR® 1500 high-strength steel has ferrite and pearlite microstructures during rolling and annealing, and the tensile strength is approximately 550 MPa. The austenite structure is obtained by heating of the sheet in a furnace to about 900 °C and holding for several minutes, and the sheet is quickly formed in the mold. After forming, holding pressure and quenching are maintained in the mold for eventually obtaining the martensite structure. The tensile strength of such martensitic steel at room temperature can reach or exceed 1500 MPa. Table 1 shows the material composition of USIBOR® 1500 high-strength steel, and Figure 1 shows the microstructure of USIBOR® 1500 high-strength steel after rolling and annealing. In the figure, α is ferrite, P is pearlite, the matrix is mainly ferrite, and the volume ratio is about 70%. The fine pearlite is uniformly distributed in the grain boundary and the crystal, and the grain size of the ferrite is less than 20 μm. The pearlite grain size is less than 2 μm.

Table 1. Material composition of USIBOR® 1500 high-strength steel.

Elements	Percentage (wt %)
C	0.2220
Mn	1.2120
P	0.0200
S	0.0032
Si	0.2590
Al	0.0362
Ti	0.0386
B	0.0038
Cr	0.1912

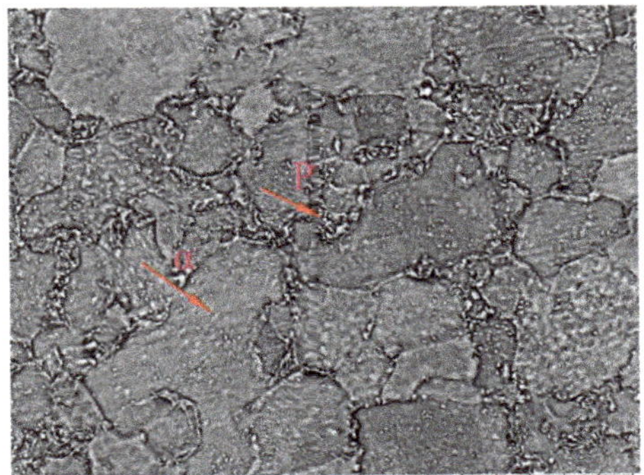

Figure 1. SEM microstructure of USIBOR® 1500 high-strength steel after rolling and annealing.

3. Results and Discussion

3.1. Phase Transformation Theory of USIBOR® 1500 High-Strength Steel

Phase transformation requires the driving force of phase change, and the resistance of phase transformation and the driving force of phase change affect the general trend of phase transformation. Through a study of the driving force of phase change, the phase equilibrium temperatures of different materials can be determined, thus judging the trend of phase transformation under various temperature conditions. When the driving force of phase transformation is zero, the corresponding temperature is the equilibrium temperature of the phase transformation. The simplified formula of the equilibrium temperature of the phase transformation is 1, 2, 3, and 4 [12]. The content of each element in the formula is the mass percentage.

$$A_{e3} = 1040 - 453C + 33Si - 27Mn - 11Cr \quad (1)$$

$$A_{e1} = 730 + 9Si - 11Mn + 6Cr, \quad (2)$$

$$B_s = 637 - 58C - 34Mn - 34Cr, \quad (3)$$

$$M_s = 539 - 423C - 7.5Si - 30Mn - 12.1Cr, \quad (4)$$

where A_{e3} is the pre-eutectoid ferrite transformation equilibrium temperature (°C), A_{e1} is the pearlite phase transformation equilibrium temperature (°C), B_s is the bainite phase transformation equilibrium temperature (°C), and M_s is the martensitic phase change equilibrium temperature (°C).

The percentage of each component in Table 1 is substituted into the calculation model of the phase equilibrium transformation temperature, and the initial phase transformation temperature of USIBOR® 1500 high-strength steel is obtained, as shown in Table 2.

Table 2. Initial phase transformation temperature of USIBOR® 1500 high-strength steel.

Phase	A_{e3}	A_{e1}	B_s	M_s
Phase change initial temperature (°C)	824	722	575	412

Through the study of martensite transformation, the transformation amount of the martensitic microstructure is only influenced by temperature but not by other factors. Therefore, the dynamic expression of the martensitic transformation is as follows [13]:

$$X = 1 - \exp[-\alpha(M_s - T)], \quad (5)$$

where X is change quantity, T is temperature (°C), M_s is martensitic transformation temperature (°C), and α is a constant that reflects the rate of martensitic transformation. The α value is 0.02 because of different organizational components.

Table 2 shows that the Ms value of USIBOR® 1500 high-strength steel is 412 °C. In order to study the microstructure transformation of USIBOR® 1500 high-strength steel, the Kirkaldy-Venugopalan model was chosen to be modified to make the model results closer to the dynamic curve of the actual microstructure transformation. The mathematical expressions of the phase transformation model are as follows [14]:

$$\dot{X} = f(G)f(T)f(C)f(X), \tag{6}$$

where $f(G)$ is the influence factor of austenite grain size, $f(T)$ is the influence factor of temperature, and $f(X)$ is the influence factor of the production transformation.

$$f(G) = 2^{AG}, \tag{7}$$

where G is austenite grain size, which can assume the value 8, and A is constant.

The austenite grain diameter D is proportional to 2G/2 and share corresponding relationships with different nucleation sites; meanwhile, the influence of austenite grains on the transformation is a mixed effect of these nucleation mechanisms in the process of continuous cold-phase transformation. Therefore, different phase transformations may have varied values. The expression of the influence of temperature is as follows:

$$f(T) = (T_{CR} - T)^n \exp(-Q/RT), \tag{8}$$

where T_{CR} is equilibrium temperature of phase change (K), R is mole constant of gas $(\mathrm{J/k \cdot mol})$, T is real-time temperature (K), Q is diffusion activation energy $(\mathrm{J/k \cdot mol})$, and n is constant. When the boundary diffuses, the value is 3. When the volume diffuses, the value is 2.

Mn and Cr in materials affect the diffusion rate of carbon atoms, and the addition of B will prolong the incubation time of ferrite transformation and reduce the ferrite transformation rate [12]. Moreover, the chemical composition of the material has different effects on different phase transformations.

Ferrite transformation:

$$f(\mathrm{C}) = (59\mathrm{Mn} + 67.7\mathrm{Cr} + 1.9 \times 10^5\mathrm{B})^{-1}, \tag{9}$$

Transition to pearlite:

$$f(\mathrm{C}) = (1.79 + 5.42\mathrm{Cr} + 3.1 \times 10^3\mathrm{B})^{-1}, \tag{10}$$

Bainite transformation:

$$f(\mathrm{C}) = ((2.34 + 10.1\mathrm{C} + 3.8\mathrm{Cr}) \times 10^{-4})^{-1}, \tag{11}$$

According to the empirical formula, the influence of the transformation quantity is expressed as follows:

$$f(\mathrm{X}) = \mathrm{X}^{0.40(1-\mathrm{X})}(1 - \mathrm{X})0.40\mathrm{X}, \tag{12}$$

According to Zener-Hillert's model [15], the transformation time and the quantity of transformation are as follows:

$$\tau(X, T) = \frac{1}{f(G)f(T)f(C)} \int_0^X \frac{dX}{f(X)}, \tag{13}$$

According to the transition temperature range of ferrite, pearlite and bainite, different temperature conditions are selected. Then the parameters are brought into the kinetic model, formula (13), to find the corresponding relationship of the time–transformation quantity under different temperature conditions. The kinetic curves of ferrite, pearlite, and bainite transformation are plotted using MATLAB (version R2016b, MathWorks, Natick, MA, USA).

The above parameters are substituted into the kinetic model, and the kinetic curves of ferrite, pearlite, and bainite transformation are plotted by MATLAB, as shown in Figure 2. The kinetic transformation curves show that the transformation rates of ferrite, pearlite, and bainite are the fastest at 580 °C, 565 °C, and 520 °C, respectively.

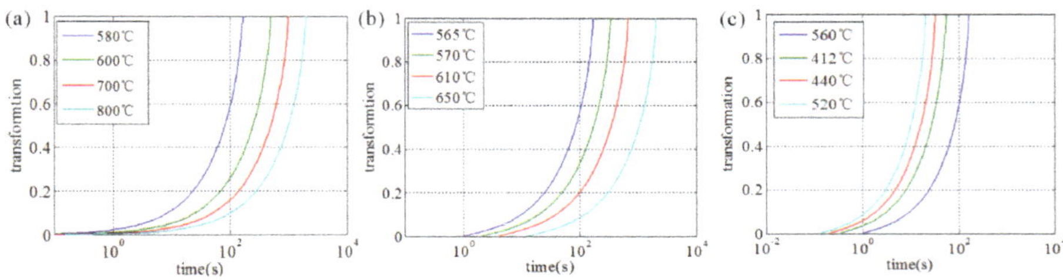

Figure 2. Dynamic curves of USIBOR® 1500 high-strength steel; (**a**) ferrite, (**b**) pearlite, (**c**) bainite.

Figure 3 depicts the time-temperature transformation (TTT) curve of USIBOR® 1500 high-strength steel; the curve is created via linear interpolation of the starting point of the dynamic curve phase transformation and the end point of the phase transformation. This curve can be used for predicting the phase transformation law of the quenching process.

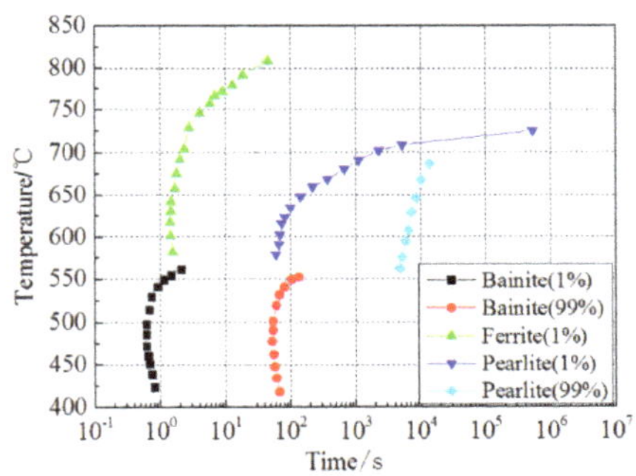

Figure 3. The time-temperature transformation (TTT) curve of USIBOR® 1500 high-strength steel.

Combined with the CCT curve drawn by Xing [16], in order to ensure the maximum martensite content of the final part, the cooling rate must be kept over 27 °C/s during the pressure-hardening quenching.

3.2. Finite Element Analysis

3.2.1. Finite Element Model and Process Scheme

Figure 4 shows the hot stamping geometry model of a vehicle anti-collision beam. The sheet used is USIBOR® 1500 high-strength steel with a thickness of 1.5 mm. Each component is meshed by DYNAFORM finite element software. The divided grids are mainly quadrilateral and account for 70% of all the grids, and triangle grids account for 30% of the total proportion. The hot stamping process scheme of USIBOR® 1500 high-strength steel is shown in Table 3.

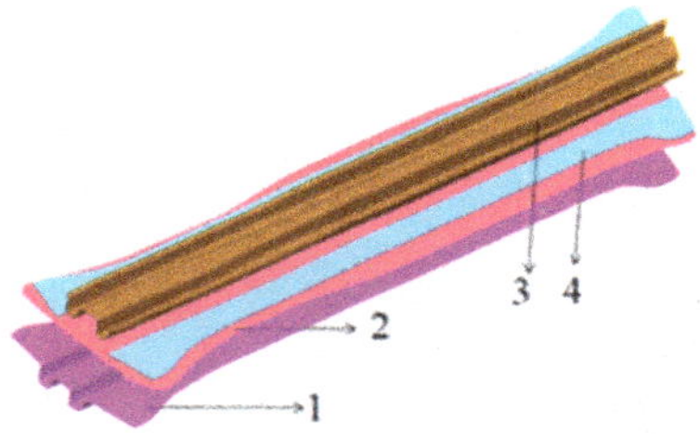

Figure 4. 3D model of hot stamping of the vehicle anti-collision beam; 1—female dye, 2—sheet, 3—punch, 4—blank holder.

Table 3. Hot stamping process scheme.

Sheet Temperature (°C)	Mould Preheating Temperature (°C)	Process Sequence
750	100	Process 1
800	100	Process 2
850	100	Process 3

3.2.2. Parameter Design

Table 4 details the specific heat capacity parameter and heat conductivity of USIBOR® 1500 high-strength steel at 20–900 °C [17]. The two parameters mentioned in the Table 4 are the most commonly used parameters in describing the thermodynamic properties of high-strength steel. The parameters of the hot stamping process of high-strength steel vary with temperature; hence, simulation must be conducted according to correct data for ensuring accuracy of the simulation results.

Table 4. Specific heat capacity and heat conductivity of USIBOR® 1500 at 20–900 °C.

Temperature (°C)	20	100	200	300	400	500	700	900
Heat conductivity $W/(m \cdot K)$	29.5	30.8	27.8	21.6	-	23.4	25.3	-
Specific heat capacity $J/(kg \cdot K)$	4.3	4.9	5.1	5.5	5.6	5.7	5.8	5.85

In the simulation analysis of hot stamping forming using DYNAFORM (version 5.9, LSTC/ETA, USA), in order to optimize the simulation effect, the appropriate material model should be selected to characterize the material. Therefore, select *MAT_ELASTIC_VISCOPLASTIC_THREMAL (#106) as the parameter setting model for the simulated material. Each parameter of the material is a variable related to temperature. Table 5 details the different parameters of USIBOR® 1500 high-strength steel with temperature at 20–900 °C [18].

Table 5. Material properties of USIBOR® 1500 high-strength steel at 20–900 °C.

Temperature (°C)	20	100	200	300	400	500	600	700	800	900
E/GPa	211	208	195	192	164	155	151	141	135	125
Poisson's ratio	0.27	0.27	0.27	0.28	0.28	0.31	0.32	0.32	0.33	0.33
Viscous parameters *C*	4.27	4.22	4.11	3.88	3.84	3.71	3.62	3.45	3.32	3.12
Viscous parameters *P*	6.2×10^9	8.4×10^5	1.4×10^4	1.3×10^3	257	80.2	41.5	30.3	21.3	31.1

According to the working conditions of hot stamping, high temperature and oxidation resistant materials must be selected. Therefore, 5CrMnMo steel is chosen as the die material. Table 6 shows the thermomechanical properties of the dye material. Table 7 depicts the specific heat capacity and heat conductivity of the dye material under 20–900 °C [19].

Table 6. Thermomechanical properties of the dye materials.

Material	Density kg/m^3	*E*/GPa	Poisson's Ratio	Specific Heat Capacity J/(kg·K)
5CrMnMo	7800	210	0.286	382

Table 7. Specific heat capacity and heat conductivity of dye material under 20–900 °C.

Temperature (°C)	20	100	200	300	400	500	600	700	800	900
Specific heat capacity $J/(kg \cdot K)$	380	388	417	448.6	481.8	528.6	540.2	542.2	540	540
Heat conductivity $W/(m \cdot K)$	62.6	83.3	84.2	74	67.13	-	57.68	-	48.21	-

In the parameter design of hot stamping simulation, the initial heating temperature of the sheet and the preheating temperature of the dye should be set when the boundary condition parameters of thermal analysis are defined. The preheating temperature of dye is 100 °C, and the initial temperatures of the sheet are 750 °C, 800 °C, and 850 °C, respectively.

The friction coefficient between the sheet and the mold during the hot stamping is set to 0.15. When the dye moves down, touches the sheet, and finally reaches the bottom dead center of the mold, contact and heat transfer must occur between the mold and the sheet. When the initial temperature of the sheet is different, the thermal contact parameters between the sheet and the mold must vary, and the contact parameters at different temperatures can be determined according to known thermodynamic conditions. Table 8 shows the thermal contact parameters at various temperatures.

Table 8. Thermal contact parameters at different temperatures.

Temperatures (°C)	400	600	700	800	900
Thermal emissivity SBC	0.092	0.098	0.101	0.102	0.113
Heat conductivity $W/(m \cdot K)$	72.32	70.23	71.45	71.70	72.21

3.2.3. Temperature Field Distribution

Figure 5a–c show that when the forming is completed, the areas with higher temperature are mainly distributed at both ends of the part and that with lower temperature are distributed on the sidewall. Meanwhile, the highest temperature is not substantially varied, and the difference in the lowest temperature is apparent. The lowest temperatures of the formed parts at the end of processes 1, 2, and 3 are 385.3 °C, 444.1 °C, and 456.9 °C, respectively.

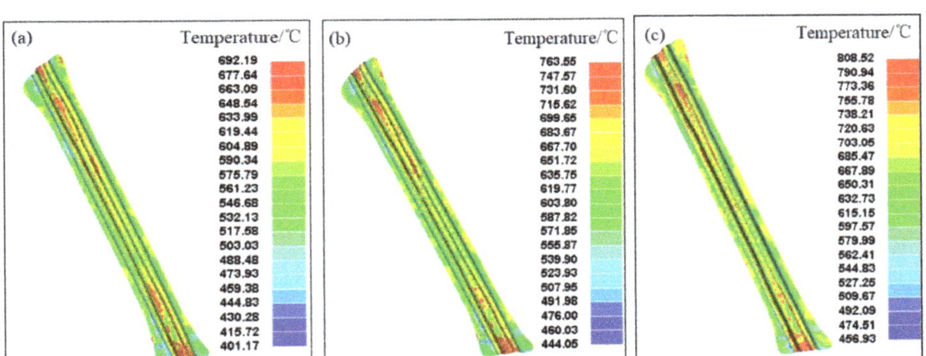

Figure 5. Temperature field distribution of formed parts at the end of hot stamping; (**a**) process 1, (**b**) process 2, and (**c**) process 3.

3.2.4. Thickness Distribution

As shown in Figure 6a–c, the smallest thickness of the formed anti-collision beam is at the two sides of the end, whereas the other parts have a more uniform thickness distribution. However, the high temperature causes the material to be softened and the formability weakened. The maximum thinning rates of the sheet at the end of processes 1, 2, and 3 are 18.7%, 26.7%, and 27.3%, respectively. These findings show that the initial temperature of the sheet has a great influence on the thickness of the final formed parts. The higher the initial temperature of the sheet, the greater the effect on the material during the forming process, thus making the thinning of the sheet increasingly evident.

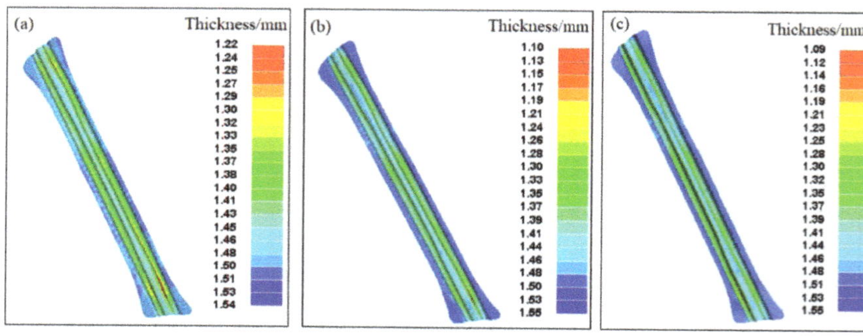

Figure 6. Thickness distribution of formed parts at end of hot stamping, (**a**) process 1, (**b**) process 2 and (**c**) process 3.

3.2.5. Martensitic Transformation Process in Quenching Process

Figure 7a–c indicates the following. No martensitic structure is found in the larger areas of the formed parts at the end of process 1; no martensitic structure is found in only smaller areas of the formed parts at the end of process 2; the position with less martensite content also reaches 87% after the quenching of process 3. These results indicate that with the increase of the initial temperature of the sheet, the martensite content of the final quenching will increase. When the initial temperature of the sheet reaches 850 °C, the martensite content exceeds 87%, and the sheet meets the requirement of 1500 MPa strength. Thus, the initial selection of the sheet temperature is 850 °C.

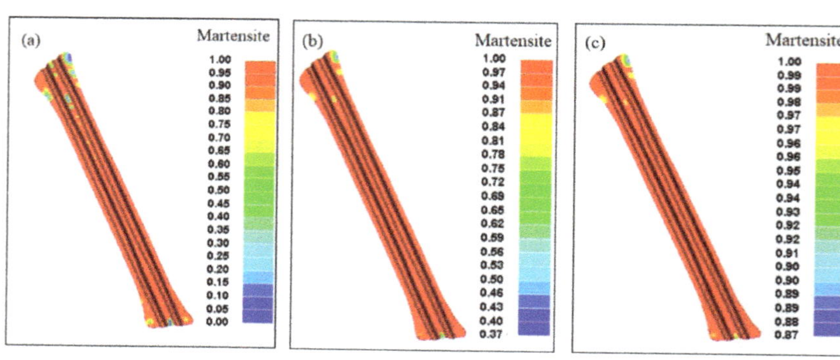

Figure 7. Martensite content of the formed part at the end of the holding pressure and quenching; (**a**) process 1, (**b**) process 2, and (**c**) process 3.

4. Conclusions

(1) A theoretical model for the phase transformation of USIBOR® 1500 high-strength steel was established. The initial phase transformation temperature of the high-strength steel phase was

calculated, and the ferrite, pearlite, and bainite transformation kinetic curves and TTT curves were plotted. An analysis of the curves showed that to ensure the martensite content of the final part, the cooling rate of quenching must be kept over 27 °C/s.

(2) The law of phase transformation of USIBOR® 1500 high-strength steel during hot stamping was studied by a simulation of the phase transformation process of the formed parts. For ensuring enhanced martensite content at the end of the final quenching, the best choice for the initial temperature of the sheet should be 850 °C.

Author Contributions: J.L. and P.Z. conceived and designed the experiments; L.Z. performed the experiments; C.X. analyzed the data; J.L. and P.Z. wrote the paper.

Funding: This research received no external funding.

Conflicts of Interest: The authors declare no conflict of interest.

References

1. Kuziak, R.; Kawalla, R.; Waengler, S. Advanced high-strength steels for automotive industry: A review. *Arch. Civ. Mech. Eng.* **2008**, *8*, 103–117. [CrossRef]
2. Matlock, D.K.; Speer, J.G. Third Generation of AHSS: Microstructure Design Concepts. In *Microstructure and Texture in Steels and Other Materials*; Haldar, A., Suwas, S., Bhattacharjee, D., Eds.; Springer: Berlin, Germany, 2009; pp. 185–205.
3. Kolleck, R.; Veit, R.; Merklein, M.; Lechler, J.; Geiger, M. Investigation on induction heating for hot stamping of boron alloyed steels. *CIRP Ann-Manuf. Technol.* **2009**, *58*, 275–278. [CrossRef]
4. Yao, Y.; Meng, J.P.; Ma, L.Y.; Zhao, G.Q.; Wang, L.R. Study on Hot Stamping and Usibor 1500P. *Appl. Mech. Mater.* **2013**, *320*, 419–425. [CrossRef]
5. Liu, W.; Liu, H.S.; Xing, Z.W.; Liu, G. Effect of tool temperature and punch speed on hot stamping of ultra high-strength steel. *Trans. Nonferrous Met. Soc. China.* **2012**, *22*, s534–s541. [CrossRef]
6. Quan, G.Z.; Wang, T.; Zhang, L. Research on the influence of hot stamping process parameters on phase field evolution by thermal-mechanical phase coupling finite element. *Int. J. Adv. Manuf. Technol.* **2016**, *89*, 1–17. [CrossRef]
7. Zhu, B.; Zhang, Y.; Li, J.; Wang, H.; Ye, Z.C. Simulation research of hot stamping and phase transition of automotive high-strength steel. *Mater. Res. Innov.* **2011**, *15*, S426–S430. [CrossRef]
8. Tamas, R.; Zoltan, F.; Imre, F. Computer simulation of steel quenching process using a multi-phase transformation model. *Comput. Mater. Sci.* **2001**, *22*, 261–278.
9. Bok, H.H.; Lee, M.G.; Kim, H.D.; Moon, M.B. Thermo-mechanical finite element simulation incorporating the temperature dependent stress-strain response of low alloy steel for practical application to the hot stamped part. *Met. Mater.-Int.* **2010**, *16*, 185–195. [CrossRef]
10. Kirkaldy, J.S.; Venugopalan, D. Prediction of microstructure and hardenability in low alloy steels. In *Phase Transformation in Ferrous Alloys, Proceedings of an International Conference, Metallurgical Soc of AIME, Ferrous Metallurgy Committee, Warrenda*; ASM, Material Science Div: Metals Park, OH, USA, 1984; pp. 125–148.
11. Bok, H.H.; Lee, M.G.; Pavlina, E.J.; Barlat, F.; Kim, H.D. Comparative study of the prediction of microstructure and mechanical properties for a hot-stamped B-pillar reinforcing part. *Int. J. Mech. Sci.* **2011**, *53*, 744–752. [CrossRef]
12. Hong, Y.Y. Research on phase transformation and numerical simulation of BR1500HS in hot stamping. Master's Thesis, Harbin Institute of Technology, Harbin, China, 1 July 2014.
13. Mori, K.; Maki, S.; Tanaka, Y. Warm and hot stamping of ultra high tensile strength steel sheets using resistance heating. *CIRP Ann-Manuf. Technol.* **2005**, *54*, 209–212. [CrossRef]
14. Hoffmann, H.; So, H.; Steinbeiss, H. Design of hot Stamping tools with cooling system. *CIRP Ann-Manuf. Technol.* **2007**, *56*, 269–272. [CrossRef]
15. Samanta, S.; Biswas, P.; Singh, S.B. Analysis of the kinetics of bainite formation below the M S, temperature. *Scr. Mater.* **2017**, *136*, 132–135. [CrossRef]
16. Hong, Y.Y.; Xing, Z.W.; Zhao, H. A model for the prediction of phase transformation of steel BR1500HS. *Adv. Commun. Technol. Syst.* **2014**, *56*, 371–378.

17. Åkerström, P.; Bergman, G.; Oldenburg, M. Numerical implementation of a constitutive model for simulation of hot stampin. *Model. Simul. Mater. Sci. Eng.* **2007**, *15*, 105–119. [CrossRef]
18. Naderi, M.; Uthaisangsuk, V.; Prahl, U.; Bleck, W. A numerical and experimental investigation into hot stamping of boron alloyed heat treated steels. *Steel Res. Int.* **2008**, *79*, 77–84. [CrossRef]
19. Naderi, M.; Durrenberger, L.; Molinari, A.; Black, W. Constitutive relationships for 22MnB5 boron steel deformed isothermally at high temperatures. *Mater. Sci. Eng. A-Struct. Mater. Prop. Microstruct. Process.* **2008**, *478*, 130–139. [CrossRef]

MDPI
St. Alban-Anlage 66
4052 Basel
Switzerland
Tel. +41 61 683 77 34
Fax +41 61 302 89 18
www.mdpi.com

Metals Editorial Office
E-mail: metals@mdpi.com
www.mdpi.com/journal/metals

www.ingramcontent.com/pod-product-compliance
Lightning Source LLC
Chambersburg PA
CBHW062053220926
51850CB00010B/2755
* 9 7 8 3 0 3 6 5 2 3 8 4 2 *